Crafting Test-Driven Software with Python

Write test suites that scale with your applications' needs and complexity using Python and PyTest

Alessandro Molina

BIRMINGHAM - MUMBAI

Crafting Test-Driven Software with Python

Copyright © 2021 Packt Publishing

Group Product Manager: Ashwin Nair
Publishing Product Manager: Ashitosh Gupta
Content Development Editor: Divya Vijayan
Senior Editor: Hayden Edwards
Technical Editor: Saurabh Kadave
Copy Editor: Safis Editing
Project Coordinator: Kinjal Bari
Proofreader: Safis Editing
Indexer: Rekha Nair
Production Designer: Prashant Ghare

First published: February 2021

Production reference: 1170221

Published by Packt Publishing Ltd.
Livery Place
35 Livery Street
Birmingham
B3 2PB, UK.

ISBN 978-1-83864-265-5

www.packt.com

To my family, Stefania, Stefano, and Cecilia, for dealing with me when I was working on this book during the evenings and weekends. To all open source contributors, for maintaining the libraries and frameworks that make our lives easier by providing us with the foundations for our daily projects.

– Alessandro Molina

Contributors

About the author

Alessandro Molina has been a Python developer since 2001, and has always been interested in Python as a web development platform. He has worked as a CTO and a team leader of Python teams for the past 10 years and is currently the core developer of the TurboGears2 web framework and the maintainer of the Beaker caching/session framework. He authored the DEPOT file storage framework and the DukPy JavaScript interpreter for Python and has collaborated on various Python projects related to web development, such as FormEncode, ToscaWidgets, and the Ming MongoDB ORM.

About the reviewer

Michael Burrows has worked for 20 years across a number of programming languages, market verticals, and software delivery roles. His main focus for the last 10 years or so has been using Python to make teams more efficient and effective.

Table of Contents

Preface

This book covers testing and test-driven development practices, introducing you to the most widespread tools and concepts that are common in the software testing community, using both the Python native `unittest` module and the `pytest` framework.

Who this book is for

This book is aimed at any Python developers that want to learn how they can test their applications and integrate testing into their development model, as well as for developers who know how to test software in other languages but are just turning to Python and thus don't yet know which tools are available to them.

What this book covers

Chapter 1, *Getting Started with Software Testing*, provides an introduction to the core concepts of automated testing and to the `unittest` Python module.

Chapter 2, *Test Doubles with a Chat Application*, presents the most common kinds of test doubles while building a real-time chat application.

Chapter 3, *Test-Driven Development while Creating a TODO List*, covers writing a todo list application adhering to the test-driven development best practices.

Chapter 4, *Scaling the Test Suite*, explores the complexities of maintaining a test suite as the software and the suite grow in size and complexity.

Chapter 5, *Introduction to PyTest*, presents the `pytest` framework and explores how it differs from the `unittest` module.

Chapter 6, *Dynamic and Parametric Tests and Fixtures*, dives into more advanced features of `pytest`, such as parametric tests and dynamic fixtures.

Chapter 7, *Fitness Function with a Contact Book Application*, dives into more advanced concepts related to acceptance tests and **Acceptance Test Driven Development (ATDD)** building a real application.

Chapter 8, *PyTest Essential Plugins*, showcases the most widespread `pytest` plugins that can be helpful in most projects.

Chapter 9, *Managing Test Environments with Tox*, presents how to manage test suites across different Python environments.

Chapter 10, *Test Documentation and Property-Based Testing*, introduces the concept of testing documentation and auto-generating tests based on the properties of the system under test.

Chapter 11, *Testing for the Web: WSGI versus HTTP*, covers how to test client-server applications based on the HTTP and WSGI protocols.

Chapter 12, *End-to-End Testing with the Robot Framework*, covers how to write tests that drive a real browser acting on a web application.

To get the most out of this book

You will need a recent version of Python 3 and pip (the package installer for Python).

All code examples in this book have been tested using Python 3.7, 3.8, and 3.9 on Linux. However, they should work on other systems too. PyTest 6.0.2 was used by the examples that rely on PyTest.

Software/hardware covered in the book	OS requirements
Python 3.7, 3.8, or 3.9	Windows, MacOSX, or Linux (any)
pip 18+	
PyTest 6.0.2+	

Additional packages and libraries will be installed from the **Python Package Index (PyPI)** using pip over the course of the chapters.

If you are using the digital version of this book, we advise you to type the code yourself or access the code via the GitHub repository (link available in the next section). Doing so will help you avoid any potential errors related to the copying and pasting of code.

Download the example code files

You can download the example code files for this book from GitHub at https://github.com/PacktPublishing/Crafting-Test-Driven-Software-with-Python. In case there's an update to the code, it will be updated on the existing GitHub repository.

We also have other code bundles from our rich catalog of books and videos available at https://github.com/PacktPublishing/. Check them out!

Conventions used

There are a number of text conventions used throughout this book.

CodeInText: Indicates code words in text, database table names, folder names, filenames, file extensions, pathnames, dummy URLs, user input, and Twitter handles. Here is an example: "The test prepares a dbpath object for the sole purpose of checking that dbmanager is asked to load that specific path."

A block of code is set as follows:

```
def test_load(self):
    dbpath = Path(tempfile.gettempdir(), "something")
    dbmanager = Mock(
        load=Mock(return_value=["buy milk", "buy water"])
    )
    app = TODOApp(io=(Mock(return_value="quit"), Mock()),
                  dbpath=dbpath, dbmanager=dbmanager)
```

When we wish to draw your attention to a particular part of a code block, the relevant lines or items are set in bold:

```
def run(self):
    self._quit = False
    while not self._quit:
        self._out(self.prompt(self.items_list()))
        command = self._in()
        self._dispatch(command)
    self._out("bye!\n")

def items_list(self):
    enumerated_items = enumerate(self._entries, start=1)
    return "\n".join(
        "{}. {}".format(idx, entry) for idx, entry in enumerated_items
    )
```

Any command-line input or output is written as follows:

```
$ pip install pytest pytest-bdd
```

Bold: Indicates a new term, an important word, or words that you see onscreen. For example, words in menus or dialog boxes appear in the text like this. Here is an example: "Select **System info** from the **Administration** panel."

 Warnings or important notes appear like this.

 Tips and tricks appear like this.

Get in touch

Feedback from our readers is always welcome.

General feedback: If you have questions about any aspect of this book, mention the book title in the subject of your message and email us at customercare@packtpub.com.

Errata: Although we have taken every care to ensure the accuracy of our content, mistakes do happen. If you have found a mistake in this book, we would be grateful if you would report this to us. Please visit www.packtpub.com/support/errata, selecting your book, clicking on the Errata Submission Form link, and entering the details.

Piracy: If you come across any illegal copies of our works in any form on the Internet, we would be grateful if you would provide us with the location address or website name. Please contact us at copyright@packt.com with a link to the material.

If you are interested in becoming an author: If there is a topic that you have expertise in and you are interested in either writing or contributing to a book, please visit authors.packtpub.com.

Reviews

Please leave a review. Once you have read and used this book, why not leave a review on the site that you purchased it from? Potential readers can then see and use your unbiased opinion to make purchase decisions, we at Packt can understand what you think about our products, and our authors can see your feedback on their book. Thank you!

For more information about Packt, please visit packt.com.

Section 1: Software Testing and Test-Driven Development

In this section, we will introduce software testing principles, automated quality control, and the distinction between quality control and quality assurance, and learn how these concepts are applied in the Python world using the pytest framework.

This section comprises the following chapters:

- Chapter 1, *Getting Started with Software Testing*
- Chapter 2, *Test Doubles with a Chat Application*
- Chapter 3, *Test-Driven Development while Creating a TODO List*
- Chapter 4, *Scaling the Test Suite*

Getting Started with Software Testing

1

Many think that the big step from "coding" to "software engineering" is made by having elegant architectures, well-defined execution plans, and software that moves big companies' processes. This mostly comes from our vision of the classic industrial product development world, where planning mostly mattered more than execution, because the execution was moved forward by an assembly line and software was an expensive internal utility that only big companies could afford

As software development science moved forward and matured, it became clear that classic industrial best practices weren't always a great fit for it. The reason being that every software product was very different, due to the technologies involved, the speed at which those technologies evolve, and in the end the fact that different software had to do totally different things. Thus the idea developed that software development was more similar to craftsmanship than to industry.

If you embrace that it's very hard, and not very effective, to try to eliminate uncertainty and issues with tons of preparation work due to the very nature of software itself, it becomes evident that the most important part of software development is detecting defects and ensuring it achieves the expected goals. Those two things are usually mostly done by having tests and a fitness function that can verify the software does what we really mean it to – founding pieces of the whole **Software Quality Control** discipline, which is what this chapter will introduce and, in practice, what this book is all about.

In this chapter, we will go through testing software products and the best practices in quality control. We will also introduce automatic tests and how they are superseding manual testing. We will take a look at what **Test-Driven Development** (TDD) is and how to apply it in Python, giving some guidance on how to distinguish between the various categories of tests, how to implement them, and how to get the right balance between test efficacy and test cost.

In this chapter, we will cover the following:

- Introducing software testing and quality control
- Introducing automatic tests and test suites
- Introducing test-driven development and unit tests
- Understanding integration and functional tests
- Understanding the testing pyramid and trophy

Technical requirements

A working Python interpreter is all that's needed.

The examples have been written in Python 3.7 but should work in most modern Python versions.

You can find the code files present in this chapter on GitHub at `https://github.com/ PacktPublishing/Crafting-Test-Driven-Software-with-Python/tree/main/Chapter01`.

Introducing software testing and quality control

From the early days, it was clear that like any other machine, software needed a way to verify it was working properly and was built with no defects.

Software development processes have been heavily inspired by manufacturing industry standards, and early on, testing and quality control were introduced into the product development life cycle. So software companies frequently have a quality assurance team that focuses on setting up processes to guarantee robust software and track results.

Those processes usually include a quality control process where the quality of the built artifact is assessed before it can be considered ready for users.

The quality control process usually achieves such confidence through the execution of a test plan. This is usually a checklist that a dedicated team goes through during the various phases of production to ensure the software behaves as expected.

Test plans

A test plan is composed of multiple **test cases**, each specifying the following:

- **Preconditions**: What's necessary to be able to verify the case
- **Steps**: Actions that have to succeed when executed in the specified order
- **Postconditions**: In which state the system is expected to be at the end of the steps

A sample test case of software where logging in with a username and password is involved, and we might want to allow the user to reset those, might look like the following table:

Test Case: 2.2 - Change User Password			
Preconditions: • A user, `user1` exists • The user is logged in as `user1` • The user is at the main menu			
#	**Action**	**Expected Response**	**Success / Fail**
1	Click the **change password** button.	The system shows a dialog to insert a new password.	
2	Enter `newpass`.	The dialog shows 7 asterisks in the password field.	
3	Click the **OK** button.	The system shows a dialog with a success message.	
4	Wait 2 seconds.	The success dialog goes away.	
Postconditions: • The `user1` password is now `newpass`			

These test cases are divided into cases, are manually verified by a dedicated team, and a sample of them is usually selected to be executed during development, but most of them are checked when the development team declared the work done.

This meant that once the team finishes its work, it takes days/weeks for the release to happen, as the whole software has to be verified by humans clicking buttons, with all the unpredictable results that involves, as humans can get distracted, pressing the wrong button or receiving phone calls in the middle of a test case.

As software usage became more widespread, and business-to-consumer products became the norm, consumers started to appreciate faster release cycles. Companies that updated their products with new features frequently were those that ended up dominating the market in the long term.

If you think about modern release cycles, we are now used to getting a new version of our favorite mobile application weekly. Such applications are probably so complex that they involve thousands of test cases. If all those cases had to be performed by a human, there would be no way for the company to provide you with frequent releases.

The worst thing you can do, by the way, is to release a broken product. Your users will lose confidence and will switch to other more reliable competitors if they can't get their job done due to crashes or bugs. So how can we deliver such frequent releases without reducing our test coverage and thus incurring more bugs?

The solution came from automating the test process. So while we learned how to detect defects by writing and executing test plans, it's only by making them automatic that we can scale them to the number of cases that will ensure robust software in the long term.

Instead of having humans test software, have some other software test it. What a person does in seconds can happen in milliseconds with software and you can run thousands of tests in a few minutes.

Introducing automatic tests and test suites

Automated testing is, in practice, the art of writing another piece of software to test an original piece of software.

As testing a whole piece of software has to take millions of variables and possible code paths into account, a single program trying to test another one would be very complex and hard to maintain. For this reason, it's usually convenient to split that program into smaller isolated programs, each being a **test case**.

Each test case contains all the instructions that are required to set up the target software in a state where the parts that are the test case areas of interest can be tested, the tests can be done, and all the conditions can be verified and reset back to the state of the target software so a subsequent test case can find a known state from which to start.

When using the unittest module that comes with the Python Standard Library, each test case is declared by subclassing from the unittest.TestCase class and adding a method whose name starts with test, which will contain the test itself:

```
import unittest

class MyTestCase(unittest.TestCase):
    def test_one(self):
        pass
```

Trying to run our previous test will do nothing by the way:

```
$ python 01_automatictests.py
$
```

We declared our test case, but we have nothing that runs it.

As for manually executed tests, the automatic tests need someone in charge of gathering all test cases and running them all. That's the role of a **test runner**.

Test runners usually involve a discovery phase (during which they detect all test cases) and a run phase (during which they run the discovered tests).

The **unittest** module provides all the components necessary to build a test runner that does both the discovery and execution of tests. For convenience, it even provides the unittest.main() method, which configures a test runner that, by default, will run the tests in the current module:

```
import unittest

class MyTestCase(unittest.TestCase):
    def test_one(self):
        pass

if __name__ == '__main__':
    unittest.main()
```

By adding a call to unittest.main() at the end of our tests, Python will automatically execute our tests when the module is invoked:

```
$ python 01_automatictests.py
.
----------------------------------------------------------------------
Ran 1 test in 0.000s

OK
```

We can confirm that the test we cared about was executed by using the -v option to print a more verbose output:

```
$ python 01_automatictests.py -v
test_one (__main__.MyTestCase) ... ok

----------------------------------------------------------------------
Ran 1 test in 0.000s

OK
```

During the discovery phase, `unittest.main` will look for all classes that inherit from `unittest.TestCase` within the module that is recognized as the main Python module (`sys.modules['__main__']`), and all those subclasses will be registered as test cases for the runner.

Individual tests are then defined by having methods with names starting with `test` in the test case classes. This means that if we add more methods with names that don't start with **test**, they won't be treated as tests:

```python
class MyTestCase(unittest.TestCase):
    def test_one(self):
        pass

    def notatest(self):
        pass
```

Trying to start the test runner again will continue to run only the `test_one` test:

```
$ python 01_automatictests.py -v
test_one (__main__.MyTestCase) ... ok

----------------------------------------------------------------------
Ran 1 test in 0.000s

OK
```

In the previous example, only the `test_one` method was executed as a test, while `notatest` was recognized as not being a test but instead as a method that we are going to use ourselves in tests.

Being able to distinguish between tests (methods whose names start with `test_`) and other methods allows us to create helpers and utility methods within our test cases that the individual tests can reuse.

Given that a test suite is a collection of multiple test cases, to grow our test suite, we need to be able to actually write more than one single `TestCase` subclass and run its tests.

Multiple test cases

We already know that `unittest.main` is the function in charge of executing our test suite, but how can we make it execute more than one `TestCase`?

The discovery phase of `unittest.main` (the phase during which `unittest.main` decides which tests to run) looks for all subclasses or `unittest.TestCase`.

The same way we had `MyTestCase` tests executed, adding more test cases is as simple as declaring more classes:

```python
import unittest


class MyTestCase(unittest.TestCase):
    def test_one(self):
        pass

    def notatest(self):
        pass


class MySecondTestCase(unittest.TestCase):
    def test_two(self):
        pass


if __name__ == '__main__':
    unittest.main()
```

Running the `01_automatictests.py` module again will lead to both test cases being verified:

```
$ python 01_automatictests.py -v
test_two (__main__.MySecondTestCase) ... ok
test_one (__main__.MyTestCase) ... ok

----------------------------------------------------------------------
Ran 2 tests in 0.000s

OK
```

If a test case is particularly complex, it can even be divided into multiple individual tests, each checking a specific subpart of it:

```python
class MySecondTestCase(unittest.TestCase):
    def test_two(self):
        pass

    def test_two_part2(self):
        pass
```

This allows us to divide the test cases into smaller pieces and eventually share setup and teardown code between the individual tests. The individual tests will be executed by the test runner in **alphabetical order**, so in this case, test_two will be executed before test_two_part2:

```
$ python 01_automatictests.py -v
test_two (__main__.MySecondTestCase) ... ok
test_two_part2 (__main__.MySecondTestCase) ... ok
test_one (__main__.MyTestCase) ... ok
```

In that run of the tests, we can see that MySecondTestCase was actually executed before MyTestCase because "MyS" is less than "MyT".

In any case, generally, it's a good idea to consider your tests as being executed in a random order and to not rely on any specific sequence of execution, because other developers might add more test cases, add more individual tests to a case, or rename classes, and you want to allow those changes with no additional issues. Especially since relying on a specific known execution order of your tests might limit your ability to parallelize your test suite and run test cases concurrently, which will be required as the size of your test suite grows.

Once more tests are added, adding them all into the same class or file quickly gets confusing, so it's usually a good idea to start organizing tests.

Organizing tests

If you have more than a few tests, it's generally a good idea to group your test cases into multiple modules and create a tests directory where you can gather the whole test plan:

```
├── 02_tests
│   ├── tests_div.py
│   └── tests_sum.py
│
```

Those tests can be executed through the `unittest discover` mode, which will look for all modules with names matching `test*.py` within a target directory and will run all the contained test cases:

```
$ python -m unittest discover 02_tests -v
test_div0 (tests_div.TestDiv) ... ok
test_div1 (tests_div.TestDiv) ... ok
test_sum0 (tests_sum.TestSum) ... ok
test_sum1 (tests_sum.TestSum) ... ok


----------------------------------------------------------------------

Ran 4 tests in 0.000s

OK
```

You can even pick which tests to run by filtering them with a substring with the `-k` parameter; for example, `-k sum` will only run tests that contain "**sum**" in their names:

```
$ python -m unittest discover 02_tests -k sum -v
test_sum0 (tests_sum.TestSum) ... ok
test_sum1 (tests_sum.TestSum) ... ok


----------------------------------------------------------------------

Ran 2 tests in 0.000s

OK
```

And yes, you can nest tests further as long as you use Python packages:

```
├──── 02_tests
│    ├──── tests_div
│    │    ├──── __init__.py
│    │    └──── tests_div.py
│    └──── tests_sum.py
```

Running tests structured like the previous directory tree will properly navigate into the subfolders and spot the nested tests.

So running `unittest` in discovery mode over that direction will properly find the `TestDiv` and `TestSum` classes declared inside the files even when they are nested in subdirectories:

```
$ python -m unittest discover 02_tests -v
test_div0 (tests_div.tests_div.TestDiv) ... ok
test_div1 (tests_div.tests_div.TestDiv) ... ok
test_sum0 (tests_sum.TestSum) ... ok
test_sum1 (tests_sum.TestSum) ... ok
```

```
------------------------------------------------------------------
Ran 4 tests in 0.000s

OK
```

Now that we know how to write tests, run them, and organize multiple tests in a test suite. We can start introducing the concept of TDD and how **unit tests** allow us to achieve it.

Introducing test-driven development and unit tests

Our tests in the previous section were all empty. The purpose was to showcase how a test suite can be made, executed, and organized in test cases and individual tests, but in the end, our tests did not test much.

Most individual tests are written following the "**Arrange, Act, Assert**" pattern:

- First, prepare any state you will need to perform the action you want to try.
- Then perform that action.
- Finally, verify the consequences of the action are those that you expected.

Generally speaking, in most cases, the action you are going to test is "calling a function," and for code that doesn't depend on any shared state, the state is usually all contained within the function arguments, so the **Arrange** phase might be omitted. Finally, the **Assert** phase will verify that the called function did what you expected, which usually means verifying the returned value and any effect at a distance that function might have:

```python
import unittest

class SomeTestCase(unittest.TestCase):
    def test_something(self):
        # Arrange phase, nothing to prepare here.

        # Act phase, call do_something
        result = do_something()

        # Assert phase, verify do_something did what we expect.
        assert result == "did something"
```

The `test_something` test is structured as a typical test with those three phases explicitly exposed, with the `do_something` call representing the **Act** phase and the final `assert` statement representing the **Assertion** phase.

Now that we know how to structure tests properly, we can see how they are helpful in implementing **TDD** and how **unit tests** are usually expressed.

Test-driven development

Tests can do more than just validating our code is doing what we expect. The TDD process argues that tests are essential in designing code itself.

Writing tests before implementing the code itself forces us to reason about our requirements. We must explicitly express requirements in a strict, well-defined way – clearly enough that a computer itself (computers are known for not being very flexible in understanding things) can understand them and state whether the code you will be writing next satisfies those requirements.

First, you write a test for your primary scenario—in this case, testing that doing 3+2 does return 5 as the result:

```python
import unittest

class AdditionTestCase(unittest.TestCase):
    def test_main(self):
        result = addition(3, 2)
        assert result == 5
```

Then you make sure it fails, which proves you are really testing something:

```
$ python 03_tdd.py
E
======================================================================
ERROR: test_main (__main__.AdditionTestCase)
----------------------------------------------------------------------
Traceback (most recent call last):
  File "03_tdd.py", line 5, in test_main
    result = addition(3, 2)
NameError: name 'addition' is not defined


----------------------------------------------------------------------
Ran 1 test in 0.000s

FAILED (errors=1)
```

Finally, you write the real code that is expected to make the test pass:

```python
def addition(arg1, arg2):
    return arg1 + arg2
```

And confirm it makes your test pass:

```
$ python 03_tdd.py
.
----------------------------------------------------------------------
Ran 1 test in 0.000s

OK
```

Once the test is done and it passes, we can revise our implementation and refactor the code. If the test still passes, it means we haven't changed the behavior and we are still doing what we wanted.

For example, we can change our addition function to unpack arguments instead of having to specify the two arguments it can receive:

```
def addition(*args):
    a1, a2 = args
    return a1 + a2
```

If our test still passes, it means we haven't changed the behavior, and it's still as good as before from that point of view:

```
$ python 03_tdd.py
.
----------------------------------------------------------------------
Ran 1 test in 0.000s

OK
```

Test-driven development is silent about **when** you reach a robust code base that satisfies all your needs. Obviously, you should at least make sure there are enough tests to cover all your requirements.

But as testing guides us in the process of development, development should guide us in the process of testing.

Looking at the code helps us come up with more white-box tests; tests that we can think of because we know how the code works internally. And while those tests might not guarantee that we are satisfying **more requirements**, they help us guarantee that our code is robust in most conditions, including corner cases.

While historically, test-first and test-driven were synonyms, today that's considered the one major difference with the **test-first** approach. In TDD we don't have the expectation to be able to write **all** tests first. Nor is it generally a good idea in the context of extreme programming practices, because you still don't know what the resulting interface that you want to test will be. What you want to test evolves as the code evolves, and we know that the code will evolve after every passing test, as a passing test gives us a chance for refactoring.

In our prior example, as we changed our `addition` function to accept a variable number of arguments, a reasonable question would be, *"But what happens if I pass three arguments? Or none?"* And our requirements, expressed by the tests, as a consequence, have to grow to support a variable number of arguments:

```
def test_threeargs(self):
    result = addition(3, 2, 1)
    assert result == 6

def test_noargs(self):
    result = addition()
    assert result == 0
```

So, writing code helped us come up with more tests to verify the conditions that came to mind when looking at the code like a white box:

```
$ python 03_tdd.py
.EE
======================================================================
ERROR: test_noargs (__main__.AdditionTestCase)
----------------------------------------------------------------------
Traceback (most recent call last):
  File "03_tdd.py", line 13, in test_noargs
    result = addition()
  File "03_tdd.py", line 18, in addition
    a1, a2 = args
ValueError: not enough values to unpack (expected 2, got 0)


======================================================================
ERROR: test_threeargs (__main__.AdditionTestCase)
----------------------------------------------------------------------
Traceback (most recent call last):
  File "03_tdd.py", line 9, in test_threeargs
    result = addition(3, 2, 1)
  File "03_tdd.py", line 18, in addition
    a1, a2 = args
ValueError: too many values to unpack (expected 2)
```

```
----------------------------------------------------------------------
Ran 3 tests in 0.001s

FAILED (errors=2)
```

And adding those failing tests helps us come up with more, and better, code that now properly handles the cases where any number of arguments is passed to our `addition` function:

```
def addition(*args):
    total = 0
    for a in args:
        total += a
    return total
```

Our `addition` function will now just iterate over the provided arguments, adding them to the total. Thus if no argument is provided, it will just return 0 because nothing was added to it.

If we run our test suite again, we will be able to confirm that both our new tests now pass, and thus we achieved what we wanted to:

```
$ python 03_tdd.py
...
----------------------------------------------------------------------
Ran 3 test in 0.001s

OK
```

Writing tests and writing code should interleave continuously. If you find yourself spending all your time on one or the other, you are probably moving away from the benefits that TDD can give you, as the two phases are meant to support each other.

There are many kinds of tests you are going to write in your test suite during your development practice, but the most common one is probably going to be test units.

Test units

The immediate question once we know how to arrange our tests, is usually "what should I test?". The answer to that is usually "it depends."

You usually want tests that assert that the feature you are providing to your users does what you expect. But do tests do nothing to guarantee that, internally, the components that collaborate with that feature behave correctly? The exposed feature might be working as a very lucky side effect of 200 different bugs in the underlying components.

So it's generally a good idea to test those units individually and verify that they all work as expected.

What are those units? Well, the answer is "it depends" again.

In most cases, you could discuss that in procedural programming, the units are the individual functions, while in object-oriented programming, it might be defined as a single class. But classes, while we usually do our best to try to isolate them to a single responsibility, might cover multiple different behaviors based on which method you call. So they actually act as multiple components in our system, and in such cases, they should be considered as separate units.

In practice, a unit is the smallest testable entity that participates in your software.

If we have a piece of software that does "multiplication," we might implement it as a main function that fetches the two provided arguments and calls a `multiply` function to do the real job:

```
def main():
    import sys
    num1, num2 = sys.argv[1:]
    num1, num2 = int(num1), int(num2)
    print(multiply(num1, num2))


def multiply(num1, num2):
    total = 0
    for _ in range(num2):
        total = addition(total, num1)
    return total


def addition(*args):
    total = 0
    for a in args:
        total += a
    return total
```

In such a case, both `addition` and `multiply` are units of our software.

While `addition` can be tested in isolation, `multiply` must use `addition` to work. `multiply` is thus defined as a **sociable unit**, while `addition` is a **solitary unit**.

Sociable unit tests are frequently also referred to as **component tests**. Your architecture mostly defines the distinction between a sociable unit test and a component test and it's hard to state exactly when one name should be preferred over the other.

While sociable units usually lead to more complete testing, they are slower, require more effort during the **Arrange** phase, and are less isolated. This means that a change in `addition` can make a test of `multiply` fail, which tells us that there is a problem, but also makes it harder to guess where the problem lies exactly.

In the subsequent chapters, we will see how sociable units can be converted into solitary units by using test doubles. If you have complete testing coverage for the underlying units, solitary unit tests can reach a level of guarantee that is similar to that of sociable units with must less effort and a faster test suite.

Test units are usually great at testing software from a white-box perspective, but that's not the sole point of view we should account for in our testing strategy. Test units guarantee that the code does what the developer meant it to, but do little to guarantee that the code does what the user needs. Integration and functional tests are usually more effective in terms of testing at that level of abstraction.

Understanding integration and functional tests

Testing all our software with solitary units can't guarantee that it's really working as expected. Unit testing confirms that the single components are working as expected, but doesn't give us any confidence about their effectiveness when paired together.

It's like testing an engine by itself, testing the wheels by themselves, testing the gears, and then expecting the car to work. We wouldn't be accounting for any issues introduced in the assembly process.

So we have a need to verify that those modules do work as expected when paired together.

That's exactly what integration tests are expected to do. They take the modules we tested individually and test them together.

Integration tests

The scope of integration tests is blurry. They might integrate two modules, or they might integrate tens of them. While they are more effective when integrating fewer modules, it's also more expensive to move forward as an approach and most developers argue that the effort of testing all possible combinations of modules in isolation isn't usually worth the benefit.

The boundary between unit tests made of sociable units and integration tests is not easy to explain. It usually depends on the architecture of the software itself. We could consider sociable units tests those tests that test units together that are inside the same architectural components, while we could consider integration tests those tests that test different architectural components together.

In an application, two separate services will be involved: `Authorization` and `Authentication`. `Authentication` takes care of letting the user in and identifying them, while `Authorization` tells us what the user can do once it is authenticated. We can see this in the following code block:

```
class Authentication:
    USERS = [{"username": "user1",
              "password": "pwd1"}]

    def login(self, username, password):
        u = self.fetch_user(username)
        if not u or u["password"] != password:
            return None
        return u

    def fetch_user(self, username):
        for u in self.USERS:
            if u["username"] == username:
                return u
        else:
            return None


class Authorization:
    PERMISSIONS = [{"user": "user1",
                    "permissions": {"create", "edit", "delete"}}]
```

```
def can(self, user, action):
    for u in self.PERMISSIONS:
        if u["user"] == user["username"]:
            return action in u["permissions"]
    else:
        return False
```

Our classes are composed of two primary methods: `Authentication.login` and `Authorization.can`. The first is in charge of authenticating the user with a username and password and returning the authenticated user, while the second is in charge of verifying that a user can do a specific action. Tests for those methods can be considered unit tests.

So `TestAuthentication.test_login` will be a unit test that verifies the behavior of the `Authentication.login` unit, while `TestAuthorization.test_can` will be a unit test that verifies the behavior of the `Authorization.can` unit:

```
class TestAuthentication(unittest.TestCase):
    def test_login(self):
        auth = Authentication()
        auth.USERS = [{"username": "testuser", "password": "testpass"}]

        resp = auth.login("testuser", "testpass")

        assert resp == {"username": "testuser", "password": "testpass"}


class TestAuthorization(unittest.TestCase):
    def test_can(self):
        authz = Authorization()
        authz.PERMISSIONS = [{"user": "testuser", "permissions":
                            {"create"}}]

        resp = authz.can({"username": "testuser"}, "create")

        assert resp is True
```

Here, we have the notable difference that `TestAuthentication.test_login` is a sociable unit test as it depends on `Authentication.fetch_user` while testing `Authentication.login`, and `TestAuthorization.test_can` is instead a solitary unit test as it doesn't depend on any other unit.

So where is the integration test?

The integration test will happen once we join those two components of our architecture (authorization and authentication) and test them together to confirm that we can actually have a user log in and verify their permissions:

```
class TestAuthorizeAuthenticatedUser(unittest.TestCase):
    def test_auth(self):
        auth = Authentication()
        authz = Authorization()
        auth.USERS = [{"username": "testuser", "password": "testpass"}]
        authz.PERMISSIONS = [{"user": "testuser",
                              "permissions": {"create"}}]

        u = auth.login("testuser", "testpass")
        resp = authz.can(u, "create")

        assert resp is True
```

Generally, it's important to be able to run your integration tests independently from your unit tests, as you will want to be able to run the unit tests continuously during development on every change:

```
$ python 05_integration.py TestAuthentication TestAuthorization
........
----------------------------------------------------------------------
Ran 8 tests in 0.000s

OK
```

While unit tests are usually verified frequently during the development cycle, it's common to run your integration tests only when you've reached a stable point where your unit tests all pass:

```
$ python 05_integration.py TestAuthorizeAuthenticatedUser
.
----------------------------------------------------------------------
Ran 1 test in 0.000s

OK
```

As you know that the units that you wrote or modified do what you expected, running the `TestAuthorizeAuthenticatedUser` case only will confirm that those entities work together as expected.

Integration tests integrate multiple components, but they actually divide themselves into many different kinds of tests depending on their purpose, with the most common kind being **functional tests**.

Functional tests

Integration tests can be very diverse. As you start integrating more and more components, you move toward a higher level of abstraction, and in the end, you move so far from the underlying components that people feel the need to distinguish those kinds of tests as they offer different benefits, complexities, and execution times.

That's why the naming of **functional** tests, **end-to-end** tests, **system** tests, **acceptance** tests, and so on all takes place.

Overall, those are all forms of integration tests; what changes are their goal and purpose:

- **Functional** tests tend to verify that we are exposing to our users the feature we actually intended. They don't care about intermediate results or side-effects; they just verify that the end result for the user is the one the specifications described, thus they are always **black-box** tests.
- **End-to-End** (E2E) tests are a specific kind of functional test that involves the vertical integration of components. The most common E2E tests are where technologies such as Selenium are involved in accessing a real application instance through a web browser.
- **System** tests are very similar to functional tests themselves, but instead of testing a single feature, they usually test a whole journey of the user across the system. So they usually simulate real usage patterns of the user to verify that the system as a whole behaves as expected.
- **Acceptance** tests are a kind of functional test that is meant to confirm that the implementation of the feature does behave as expected. They usually express the primary usage flow of the feature, leaving less common flows for other integration tests, and are frequently provided by the specifications themselves to help the developer confirm that they implemented what was expected.

But those are not the only kinds of integration that people refer to; new types are continuously defined in the effort to distinguish the goals of tests and responsibilities. **Component** tests, **contract** tests, and many others are kinds of tests whose goal is to verify integration between different pieces of the software at different layers. Overall, you shouldn't be ashamed of asking your colleagues what they mean exactly when they use those names, because you will notice each one of them will value different properties of those tests when classifying them into the different categories.

The general distinction to keep in mind when distinguishing between integration tests and functional tests is that unit and integration tests aim to test the implementation, while functional tests aim to test the behavior.

How you do that can easily involve the same exact technologies and it's just a matter of different goals. Properly covering the behavior of your software with the right kind of tests can be the difference between buggy software and reliable software. That's why there has been a long debate about how to structure test suites, leading to the testing pyramid and the testing trophy as the most widespread models of test distribution.

Understanding the testing pyramid and trophy

Given the need to provide different kinds of tests – unit, integration, and E2E as each one of them has different benefits and costs, the next immediate question is how do we get the right balance?

Each kind of test comes with a benefit and a cost, so it's a matter of finding where we get the best return on investment:

- **E2E** tests verify the real experience of what the user faces. They are, in theory, the most realistic kind of tests and can detect problems such as incompatibilities with specific platforms (for example, browsers) and exercise our system as a whole. But when something goes wrong, it is hard to spot where the problem lies. They are very slow and tend to be flaky (failing for reasons unrelated to our software, such as network conditions).
- **Integration** tests usually provide a reasonable guarantee that the software is doing what it is expected to do and are fairly robust to internal implementation changes, requiring less frequent refactoring when the internals of the software change. But they can still get very slow if your system involves writes to database services, the rendering of page templates, routing HTTP requests, and generally slow parts. And when something goes wrong, we might have to go through tens of layers before being able to spot where the problem is.
- **Unit** tests can be very fast (especially when talking of solitary units) and provide very pinpointed information about where problems are. But they can't always guarantee that the software as a whole does what it's expected to do and can make changing implementation details expensive because a change to internals that don't impact the software behavior might require changing tens of unit tests.

Each of them has its own pros and cons, and the development community has long argued how to get the right balance.

The two primary models that have emerged are the testing pyramid and the testing trophy, named after their shapes.

The testing pyramid

The testing pyramid originates from Mike Cohn's *Succeeding with Agile* book, where the two rules of thumb are "*Write test with different granularities*" (so you should have unit, integration, E2E, and so on...) and "*the more you get high level, the less you should test*" (so you should have tons of unit tests, and a few E2E tests).

While different people will argue about which different layers are contained within it, the testing pyramid can be simplified to look like this:

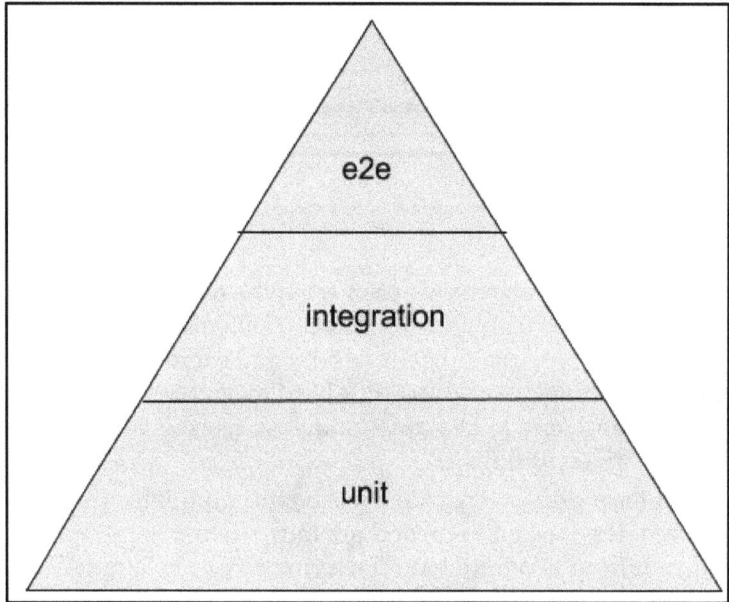

Figure 1.1 – Testing pyramid

The tip of the pyramid is narrow, thus meaning we have fewer of those tests, while the base is wider, meaning we should mostly cover code with those kinds of tests. So, as we move down through the layers, the lower we get, the more tests we should have.

The idea is that as unit tests are fast to run and expose pinpointed issues early on, you should have a lot of them and shrink the number of tests as they move to higher layers and thus get slower and vaguer about what's broken.

The testing pyramid is probably the most widespread practice for organizing tests and usually pairs well with test-driven development as unit tests are the founding tool for the TDD process.

The other most widespread model is the testing trophy, which instead emphasizes integration tests.

The testing trophy

The testing trophy originates from a phrase by Guillermo Rauch, the author of Socket.io and many other famous JavaScript-based technologies. Guillermo stated that developers should "*Write tests. Not too many. Mostly integration.*"

Like Mike Cohn, he clearly states that tests are the foundation of any effective software development practice, but he argues that they have a diminishing return and thus it's important to find the sweet spot where you get the best return on the time spent writing tests.

That sweet spot is expected to live in integration tests because you usually need fewer of them to spot real problems, they are not too bound to implementation details, and they are still fast enough that you can afford to write a few of them.

So the testing trophy will look like this:

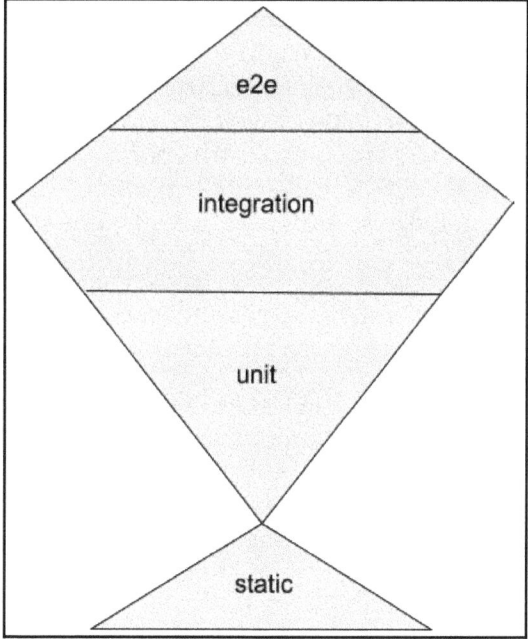

Figure 1.2 – Testing trophy

As you probably saw, the testing trophy puts a lot of value on **static** tests too, because the whole idea of the testing trophy is that what is really of value is the return on investment, and static checks are fairly cheap, up to the point that most development environments run them in real time. Linters, type checkers, and more advanced kinds of type analyzers are cheap enough that it would do no good to ignore them even if they are rarely able to spot bugs in your business logic.

Unit tests instead can cost developers time with the need to adapt them due to internal implementation detail changes that don't impact the final behavior of the software in any way, and thus the effort spent on them should be kept under control.

Those two models are the most common ways to distribute your tests, but more best practices are involved when thinking of testing distribution and coverage.

Testing distribution and coverage

While the importance of testing is widely recognized, there is also general agreement that test suites have a diminishing return.

There is little point in wasting hours on testing plain getters and setters or testing internal/private methods. The sweet spot is said to be around 80% of code coverage, even though I think that really depends on the language in use – the more expressive your language is, the less code you have to write to perform complex actions. And all complex actions should be properly tested, so in the case of Python, the sweet spots probably lies more in the range of 90%. But there are cases, such as porting projects from Python 2 to Python 3, where code coverage of 100% is the only way you can confirm that you haven't changed any behavior at all in the process of porting your code base.

Last but not least, most testing practices related to test-driven development take care of the testing practice up to the release point. It's important to keep in mind that when the software is released, the testing process hasn't finished.

Many teams forget to set up proper **system tests** and don't have a way to identify and reproduce issues that can only happen in production environments with real concurrent users and large amounts of data. Having staging environments and a suite to simulate incidents or real users' behaviors might be the only way to spot bugs that only happen after days of continuous use of the system. And some companies go as far as testing the production system with tools that inject real problems continuously for the sole purpose of verifying that the system is solid.

Summary

As we saw in the sections about integration tests, functional tests, and the testing pyramid/trophy models, there are many different visions about what should be tested, with which goals in mind, and how test suites should be organized. Getting this right can impact how much you trust your automatic test suite, and thus how much you evolve it because it provides you with value.

Learning to do proper automated testing is the gateway to major software development boosts, opening possibilities for practices such as continuous integration and continuous delivery, which would otherwise be impossible without a proper test suite.

But testing isn't easy; it comes with many side-effects that are not immediately obvious, and for which the software development industry started to provide tools and best practices only recently. So in the next chapters, we will look at some of those best practices and tools that can help you write a good, easily maintained test suite.

2
Test Doubles with a Chat Application

We have seen how a test suite, to be reasonably reliable, should include various kinds of tests that cover components at various levels. Usually, tests, in regard to how many components they involve, are categorized into at least three kinds: unit, integration, and end-to-end.

Test doubles ease the implementation of tests by breaking dependencies between components and allowing us to simulate the behaviors we want.

In this chapter, we will look at the most common kinds of test doubles, what their goals are, and how to use them in real code. By the end of this chapter, we will have covered how to use all those test doubles and you will be able to leverage them for your own Python projects.

By adding test doubles to your toolchain, you will be able to write faster tests, decouple the components you want to test from the rest of the system, simulate behaviors that depend on other components' state, and in general move your test suite development forward with fewer blockers.

In this chapter, we will learn how to move forward, in the **Test-Driven Development (TDD)** way, the development of an application that depends on other external dependencies such as a database management system and networking, relying on test doubles for the development process and replacing them in our inner test layers to ensure fast and consistent execution of our tests.

In this chapter, we will cover the following topics:

- Introducing test doubles
- Starting our chat application with TDD
- Using dummy objects
- Replacing components with stubs

- Checking behaviors with spies
- Using mocks
- Replacing dependencies with fakes
- Understanding acceptance tests and doubles
- Managing dependencies with dependency injection

Technical requirements

A working Python interpreter should be all that is needed.

The examples have been written on Python 3.7 but should work on most modern Python versions.

You can find the code files present in this chapter on GitHub at `https://github.com/ PacktPublishing/Crafting-Test-Driven-Software-with-Python/tree/main/Chapter02`.

Introducing test doubles

In test-driven development, the tests drive the development process and architecture. The software design evolves as the software changes during the development of new tests, and the architecture you end up with should be a consequence of the need to satisfy your tests.

Tests are thus the arbiter that decides the future of our software and declares that the software is doing what it is designed for. There are specific kinds of tests that are explicitly designed to tell us that the software is doing what it was requested: Acceptance and Functional tests.

So, while there are two possible approaches to TDD, top-down and bottom-up (one starting with higher-level tests first, and the other starting with unit tests first), the best way to avoid going in the wrong direction is to always keep in mind your acceptance rules, and the most effective way to do so is to write them down as tests.

But how can we write a test that depends on the whole software existing and working if we haven't yet written the software at all? The key is **test doubles**: objects that are able to replace missing, incomplete, or expensive parts of our code just for the purpose of testing.

A test double is an object that takes the place of another object, faking that it is actually able to do the same things as the other object, while in reality, it does nothing.

But if we make our tests pass with test doubles, how do we avoid shipping software that is just a bunch of fake entities? That's why it's important to have various layers of tests – the more you move up through the layers, the fewer test doubles you should have, all the way up to end-to-end tests, which should involve no test doubles at all.

Test-driven development also suggests that we should write the minimum amount of code necessary to make a test pass and it's a very important rule because, otherwise, you could easily end up writing code whose development has to be driven by other new tests.

That means that to have a fairly high-level test (such as an acceptance test) pass, we are probably going to involve many test doubles at the beginning (as our software is still empty). So when are we expected to replace those test doubles with real objects?

That's where *Test-Driven Development by Example* by *Kent Beck* suggests relying on a TODO list. As you write your code, you should write down anything that you think you need to improve/support/replace. And before moving forward to writing the next acceptance test, the TODO list should be completed.

In your TODO list, you can record entries to replace the test doubles with real objects. As a consequence, we are going to write tests that verify the behaviors of those real objects and, subsequently, their implementation, finally replacing them with the real objects themselves in our original acceptance test to confirm it still passes.

To showcase how test doubles can help us during TDD, we are going to build a chat application by relying on the most common kind of test doubles.

Starting our chat application with TDD

When you start the development of a new feature, the first test you might want to write is the primary acceptance test – the one that helps you define "this is what I want to achieve." Acceptance tests expose the components we need to create and the behaviors they need to have, allowing us to move forward by designing the development tests for those components and thus writing down unit and integration tests.

In the case of the chat application, our acceptance test will probably be a test where one user can send a message and another user can receive it:

```python
import unittest

class TestChatAcceptance(unittest.TestCase):
    def test_message_exchange(self):
        user1 = ChatClient("John Doe")
        user2 = ChatClient("Harry Potter")
```

```
            user1.send_message("Hello World")
            messages = user2.fetch_messages()
            assert messages == ["John Doe: Hello World"]


    if __name__ == '__main__':
        unittest.main()
```

Our test makes clear that we want two ChatClient instances that exchange a message. The first sends a new message and the second is able to fetch it and see it.

Now, we made our mind up about the fact that we want two chat clients to exist: one that sends messages and another that can receive them. We will surely evolve this simple vision of our application in the future, but so far it has helped us set some clear expectations.

The ChatClient class doesn't yet exist, by the way. We vaguely know that we want it to be able to send messages and fetch messages, but we still lack tons of details about what it should do and how it should do it. So the next step is to start clarifying what we want those capabilities to look like.

If we run our acceptance test, by running the 01_chat_acceptance.py file where we saved the previous test case, it will fail with an error:

```
$ python 01_chat_acceptance.py TestChatAcceptance
E
======================================================================
ERROR: test_message_exchange (__main__.TestChatAcceptance)
----------------------------------------------------------------------
Traceback (most recent call last):
  File "01_chat_acceptance.py", line 5, in test_message_exchange
    user1 = ChatClient("John Doe")
NameError: global name 'ChatClient' is not defined


----------------------------------------------------------------------
Ran 1 test in 0.000s

FAILED (errors=1)
```

By complaining that ChatClient is not defined, it will point out that our next step should probably be writing a client.

So we know that the first thing we have to start with is creating a `ChatClient` and, as we want that client to be able to remember the nickname of the user, we need to ensure that it's aware of the nickname of the user. So let's start by writing a development test to ensure that `ChatClient` will able to do so:

```
class TestChatClient(unittest.TestCase):
    def test_nickname(self):
        client = ChatClient("User 1")

        assert client.nickname == "User 1"
```

At this point, we already know that both our acceptance test and our development test will fail, as we haven't yet written any implementation. But let's confirm our test suite does what we expect:

```
$ python 01_chat_acceptance.py TestChatClient
E
========================================================================
ERROR: test_nickname (__main__.TestChatClient)
------------------------------------------------------------------------
Traceback (most recent call last):
  File "01_chat_acceptance.py", line 16, in test_nickname
    client = ChatClient("User 1")
NameError: global name 'ChatClient' is not defined


------------------------------------------------------------------------
Ran 1 test in 0.000s

FAILED (errors=1)
```

Obviously, our test is failing with the fact that `ChatClient` doesn't even exist, so let's implement the `ChatClient` class itself and make it aware of the nickname used:

```
class ChatClient:
    def __init__(self, nickname):
        self.nickname = nickname
```

Now, rerunning our test unit should be successful, as we created the `ChatClient` and we made it able to keep the memory of the user's nickname that is connected to our chat application:

```
$ python 01_dummy.py TestChatClient
.
------------------------------------------------------------------------
Ran 1 test in 0.000s

OK
```

So our unit test now passes, and we can move forward. What needs to be done next? To know that, we just have to go back and run our acceptance test again. Does it pass? Does it need any other unit to be developed?

```
$ python 01_chat_acceptance.py TestChatAcceptance
E
================================================================
ERROR: test_message_exchange (__main__.TestChatAcceptance)
----------------------------------------------------------------
Traceback (most recent call last):
  File "01_chat_acceptance.py", line 8, in test_message_exchange
    user1.send_message("Hello World")
AttributeError: ChatClient instance has no attribute 'send_message'


----------------------------------------------------------------
Ran 1 test in 0.000s

FAILED (errors=1)
```

Running our acceptance test again, it has now complained about the `ChatClient.send_message` method, so now we know that we need to work on that unit next. As is usually expected with a TDD approach, we can start the work with a unit test.

So let's extend our `TestChatClient` case with one additional `test_send_message` test:

```python
class TestChatClient(unittest.TestCase):
    def test_nickname(self):
        client = ChatClient("User 1")

        assert client.nickname == "User 1"

    def test_send_message(self):
        client = ChatClient("User 1")
        sent_message = client.send_message("Hello World")
        assert sent_message == "User 1: Hello World"
```

The new `test_send_message` test creates a client for `User 1` and then sends a message to the chat from that user, verifying that the outgoing message was actually submitted as a message sent by that user.

Going back to our shell and rerunning our tests for the `ChatClient` component will tell us that we now have to write that method to satisfy the test:

```
$ python 01_chat_acceptance.py TestChatClient
.E
================================================================
ERROR: test_send_message (__main__.TestChatClient)
```

```
  ----------------------------------------------------------------------
  Traceback (most recent call last):
    File "01_chat_acceptance.py", line 22, in test_send_message
      sent_message = client.send_message("Hello World")
  AttributeError: ChatClient instance has no attribute 'send_message'

  ----------------------------------------------------------------------
  Ran 2 tests in 0.000s

  FAILED (errors=1)
```

So let's move back to development again and add the `send_message` method to our component. We already decided that it has to accept the message, prefix it with the sender's nickname, and probably send it to all the other users:

```
class ChatClient:
    def __init__(self, nickname):
        self.nickname = nickname

    def send_message(self, message):
        sent_message = "{}: {}".format(self.nickname, message)
        self.connection.broadcast(message)
        return sent_message
```

Let's rerun our test case for the component to confirm that we now satisfy it:

```
$ python 01_chat_acceptance.py TestChatClient
.E
======================================================================
ERROR: test_send_message (__main__.TestChatClient)
----------------------------------------------------------------------
Traceback (most recent call last):
  File "01_chat_acceptance.py", line 22, in test_send_message
    sent_message = client.send_message("Hello World")
  File "01_chat_acceptance.py", line 32, in send_message
    self.connection.broadcast(message)
AttributeError: ChatClient instance has no attribute 'connection'

----------------------------------------------------------------------
Ran 2 tests in 0.000s

FAILED (errors=1)
```

Our test failed again – it told us that our `ChatClient.send_message` method is now there, and the test was able to call it, but it's not yet working.

This is because we actually went a bit further in having the client already send the messages through the network before the need was exposed by our tests. But we already knew that's what we actually wanted to do anyway, and it actually serves the purpose of introducing our first test double: **dummy objects**.

Using dummy objects

A dummy is an object that does nothing. It just serves the purpose of being passed around as an argument and not making the code crash because we lack an object. But its implementation is totally empty; it does nothing.

In our chat application, we need a connection object to be able to send messages from one client to the other. We have not yet implemented that connection object, and for now, we are focused on having the `ChatClient.send_message` test pass, but how can we make it pass if we don't yet have a working `Connection` object the client relies on?

That's where **dummy objects** come in handy. They replace other objects, faking that they can do their job, but in reality, they do absolutely nothing.

A dummy object for our `Connection` class would currently look like this:

```
class _DummyConnection:
    def broadcast(*args, **kwargs):
        pass
```

In practice, it's an object that provides a `broadcast` method but does absolutely nothing. Dummy objects are just fillers for the arguments of properties that another object needs. They are frequently not even used at all and just provide a pass-through to satisfy some required argument.

Now we can adapt our previous `TestChatClient.test_send_message` test to use a dummy for the connection by setting `client.connection` to `_DummyConnection`. That should make our test pass as we broke the dependency over a real connection:

```
class TestChatClient(unittest.TestCase):
    ...

    def test_send_message(self):
        client = ChatClient("User 1")
        client.connection = _DummyConnection()
```

```
    sent_message = client.send_message("Hello World")

    assert sent_message == "User 1: Hello World"
```

Another convenient way to implement dummy objects is just to use the Python
unittest.mock module. In the *Using mocks* section, will see that, while the name is pretty
specific, the unittest.mock.Mock object is in practice able to serve all test doubles cases
introduced in this chapter. It just depends on which features we use and which we ignore.

So in our previous example, we can just replace our _DummyConnection
with unittest.mock.Mock and avoid having to implement a dedicated class at all:

```python
import unittest.mock

class TestChatClient(unittest.TestCase):
    ...

    def test_send_message(self):
        client = ChatClient("User 1")
        client.connection = unittest.mock.Mock()
        sent_message = client.send_message("Hello World")

        assert sent_message == "User 1: Hello World"
```

If we run our tests again for TestChatClient, we should see that we finally succeeded in
making them pass:

```
$ python 02_chat_dummy.py TestChatClient
..
----------------------------------------------------------------------
Ran 2 tests in 0.000s

OK
```

Does that mean that our work is done? Not yet, because checking our acceptance test
(TestChatAcceptance) again will tell us that we are not yet there:

```
$ python 02_chat_dummy.py TestChatAcceptance
E
======================================================================
ERROR: test_message_exchange (__main__.TestChatAcceptance)
----------------------------------------------------------------------
Traceback (most recent call last):
  File "02_chat_dummy.py", line 8, in test_message_exchange
    user1.send_message("Hello World")
  File "02_chat_dummy.py", line 39, in send_message
    self.connection.broadcast(message)
```

```
AttributeError: ChatClient instance has no attribute 'connection'

-------------------------------------------------------------------
Ran 1 test in 0.000s

FAILED (errors=1)
```

We implemented the `ChatClient.send_message` method and it passes its test, but our acceptance test is now reminding us that we still have to implement the `Connection` object as we just used a double for it in the `send_message` test.

The connection object is the next thing we are going to implement, but the connection will need to be able to reach a server that can route the messages to all connected clients, and making our tests pass a `DummyConnection` won't be enough anymore. We will have to actually see the messages and thus using **stubs** will be necessary.

Replacing components with stubs

Our `connection` object will be in charge of making our message available to all the other clients and, probably in the near future, letting us know when there are new messages.

The first step to drive the development of our `Connection` object is to start building a `TestConnection` test case and a `test_broadcast` test to make our expectations of the implementation clear:

```python
class TestConnection(unittest.TestCase):
    def test_broadcast(self):
        c = Connection(("localhost", 9090))

        c.broadcast("some message")

        assert c.get_messages()[-1] == "some message"
```

Our test specifies that once we've sent a message in broadcast, the latest entry in the messages visible in the chat should be our own message (as it was the last message sent). Obviously, running our test now will fail because the `Connection` object doesn't exist at all, so let's make one.

A possible idea for how to implement cross-client communication is to use a `multiprocessing.managers.SyncManager` and store the messages in a list that is accessible by all the clients that connect to it.

The only thing we will have to do is register a single `Connection.get_messages` identifier in the manager. The purpose of that identifier will be to return the list of messages that are currently in the chat so that `ChatClient` can read them or append new messages:

```
from multiprocessing.managers import SyncManager

class Connection(SyncManager):
    def __init__(self, address):
        self.register("get_messages")
        super().__init__(address=address, authkey=b'mychatsecret')
        self.connect()
```

Then the `Connection.broadcast` method will be as simple as just getting the messages through `Connection.get_messages` and appending a new message to them:

```
from multiprocessing.managers import SyncManager

class Connection(SyncManager):
    def __init__(self, address):
        self.register("get_messages")
        super().__init__(address=address, authkey=b'mychatsecret')
        self.connect()

    def broadcast(self, message):
        messages = self.get_messages()
        messages.append(message)
```

Now our `connection` object is done and it provides a `broadcast` method, so we can verify that it does add a new message to our chat by rerunning our test:

```
$ python 03_chat_stubs.py TestConnection
E
======================================================================
ERROR: test_broadcast (__main__.TestConnection)
----------------------------------------------------------------------
Traceback (most recent call last):
  File "03_chat_stubs.py", line 33, in test_broadcast
    c = Connection(("localhost", 9090))
  File "03_chat_stubs.py", line 56, in __init__
    self.connect()
  File "/usr/lib/python3.7/multiprocessing/managers.py", line 532, in
connect
    conn = Client(self._address, authkey=self._authkey)
  File "/usr/lib/python3.7/multiprocessing/connection.py", line 492, in
Client
    c = SocketClient(address)
```

```
   File "/usr/lib/python3.7/multiprocessing/connection.py", line 619, in
SocketClient
      s.connect(address)
ConnectionRefusedError: [Errno 111] Connection refused


----------------------------------------------------------------------
Ran 1 test in 0.003s

FAILED (errors=1)
```

Sadly, our test failed, because we still don't have a server, so the connection couldn't get created because there is no server it could connect to. Until we have a server, we already know we can replace our `Connection.connect` method with a dummy in our test and retry:

```python
class TestConnection(unittest.TestCase):
    def test_broadcast(self):
        with unittest.mock.patch.object(Connection, "connect"):
            c = Connection(("localhost", 9090))

        c.broadcast("some message")

        assert c.get_messages()[-1] == "some message"
```

`unittest.mock.patch.object` is a convenience method that allows us to replace a method or attribute of an object with a `unittest.mock.Mock` for the whole duration of the code block within the context. So in this case, we disabled the `Connection.connect` method so that the connection could be created without a server.

Okay, so now we expect our test to finally pass, right? Let's try to run it once more:

```
$ python 03_chat_stubs.py TestConnection
F
======================================================================
FAIL: test_broadcast (__main__.TestConnection)
----------------------------------------------------------------------
Traceback (most recent call last):
  File "03_chat_stubs.py", line 36, in test_broadcast
    c.broadcast("some message")
  File "03_chat_stubs.py", line 60, in broadcast
    messages = self.get_messages()
  File "/usr/lib/python3.7/multiprocessing/managers.py", line 724, in temp
    token, exp = self._create(typeid, *args, **kwds)
  File "/usr/lib/python3.7/multiprocessing/managers.py", line 606, in
_create
    assert self._state.value == State.STARTED, 'server not yet started'
AssertionError: server not yet started
```

```
---------------------------------------------------------------------
Ran 1 test in 0.001s

FAILED (failures=1)
```

Not really. The object was successfully created, but once we tried to get the chat messages to append the new one, it failed, as there was no server we could connect to.

But we do really need the messages, otherwise, the whole test has no way to verify that the message was added to the existing messages and thus sent. So what can we do?

Here is where stubs come in handy. Stubs provide canned answers, replacing those pieces of the software with the ready-made state or answer that we could have got if it had run for real. So we are going to replace Connection.get_messages with a stub that returns an empty list and see that everything works as we expected:

```python
class TestConnection(unittest.TestCase):
    def test_broadcast(self):
        with unittest.mock.patch.object(Connection, "connect"):
            c = Connection(("localhost", 9090))

            with unittest.mock.patch.object(c, "get_messages",
                                            return_value=[]):
                c.broadcast("some message")

                assert c.get_messages()[-1] == "some message"
```

You can now see that, after the first call to unittest.mock.patch.object, where we replaced the connect method with a dummy one, we now have a second one. In this one, we replace the get_messages method of the newly made Connection instance with one that returns a canned response of an empty list, simulating this being the first message that was sent to the chat.

Finally, running our tests again will confirm that the Connection.broadcast method is doing what we expected:

```
$ python 03_chat_stubs.py TestConnection
.
---------------------------------------------------------------------
Ran 1 test in 0.001s
OK
```

Okay, so now we have ChatClient and Connection tests passing, so we clearly did our job, right?

Let's check whether that's true by running our acceptance test:

```
$ python 03_chat_stubs.py TestChatAcceptance
E
======================================================================
ERROR: test_message_exchange (__main__.TestChatAcceptance)
----------------------------------------------------------------------
Traceback (most recent call last):
  File "03_chat_stubs.py", line 10, in test_message_exchange
    user1.send_message("Hello World")
  File "03_chat_stubs.py", line 49, in send_message
    self.connection.broadcast(message)
AttributeError: 'ChatClient' object has no attribute 'connection'


----------------------------------------------------------------------
Ran 1 test in 0.000s

FAILED (errors=1)
```

Not yet... We made the `Connection` object, but we clearly forgot to bind it to our `ChatClient`.

So let's move forward by binding `ChatClient` with `Connection` and introducing spies as a way to verify that the `ChatClient` is using the `Connection` the way we actually expect it to.

Checking behaviors with spies

We know that `ChatClient` must use a connection to send and receive the messages. So the next thing we have to do to make sure our `test_message_exchange` test passes is to make sure that the connection exists and is used. But we don't want to establish a connection every time a `ChatClient` is created, so the idea is to create a connection through a method that lazily makes them when they're needed the first time.

We will call this method `ChatClient._get_connection` and we want to make sure that `ChatClient` will actually use the connection provided by that method. To verify that `ChatClient` uses the provided connection, we are going to set up a test with a **spy**, a kind of dummy object that, instead of doing nothing, actually records how it was called (if it was) and with which arguments.

As we did when setting up the stub, we are going to use `unittest.mock.patch` to replace the `ChatClient._get_connection` method with a stub that, instead of returning the connection, returns the spy. Then we are going to check through the spy that the `ChatClient.send_message` method actually used the connection we returned to send the message:

```
def test_client_connection(self):
    client = ChatClient("User 1")

    connection_spy = unittest.mock.MagicMock()
    with unittest.mock.patch.object(client, "_get_connection",
                                    return_value=connection_spy):
        client.send_message("Hello World")

    # assert that the spy was called with the
    # expected data to broadcast.
    connection_spy.broadcast.assert_called_with(("User 1:
                                                Hello World"))
```

Now if we call our test, it's going to fail because we never made a `ChatClient._get_connection` method and thus it can't be replaced with the stub:

```
$ python 04_chat_spies.py TestChatClient
E..
========================================================================
ERROR: test_client_connection (__main__.TestChatClient)
------------------------------------------------------------------------
Traceback (most recent call last):
  File "04_chat_spies.py", line 35, in test_client_connection
    return_value=connection_spy):
  File "/usr/lib/python3.7/unittest/mock.py", line 1319, in __enter__
    original, local = self.get_original()
  File "/usr/lib/python3.7/unittest/mock.py", line 1293, in get_original
    "%s does not have the attribute %r" % (target, name)
AttributeError: <__main__.ChatClient object at 0x7f962dd8d050> does not
have the attribute '_get_connection'

------------------------------------------------------------------------
Ran 3 tests in 0.001s

FAILED (errors=1)
```

So let's go to our `ChatClient` class and let's add the `_get_connection` method, which is going to return a new `Connection` object against a predefined port where the server will listen locally (normally, we would make the port and host for a service configurable, but given that it's just a simple chat application for our own use, we can take for granted that the server will run on a known port and host):

```python
class ChatClient:
    def __init__(self, nickname):
        self.nickname = nickname

    def send_message(self, message):
        sent_message = "{}: {}".format(self.nickname, message)
        self.connection.broadcast(message)
        return sent_message

    def _get_connection(self):
        return Connection(("localhost", 9090))
```

Great – so our test should be happy now! The stub can be put in place, so let's see what happens when running our tests again:

```
$ python 04_chat_spies.py TestChatClient
E..
======================================================================
ERROR: test_client_connection (__main__.TestChatClient)
----------------------------------------------------------------------
Traceback (most recent call last):
  File "04_chat_spies.py", line 36, in test_client_connection
    client.send_message("Hello World")
  File "04_chat_spies.py", line 83, in send_message
    self.connection.broadcast(message)
AttributeError: 'ChatClient' object has no attribute 'connection'

----------------------------------------------------------------------
Ran 3 tests in 0.001s

FAILED (errors=1)
```

Okay, we made the `_get_connection` but the `ChatClient` never calls it... So the object is still missing a `connection` attribute.

We know we want this attribute to lazily create the connection, so we are going to define a property that calls `_get_connection` the first time it's accessed:

```python
class ChatClient:
    def __init__(self, nickname):
        self.nickname = nickname
```

```
            self._connection = None

    def send_message(self, message):
        sent_message = "{}: {}".format(self.nickname, message)
        self.connection.broadcast(message)
        return sent_message

    @property
    def connection(self):
        if self._connection is None:
            self._connection = self._get_connection()
        return self._connection

    @connection.setter
    def connection(self, value):
        if self._connection is not None:
            self._connection.close()
        self._connection = value

    def _get_connection(self):
        return Connection(("localhost", 9090))
```

Now when `ChatClient.connection` is accessed, as `ChatClient._connection` will be `None`, the `ChatClient._get_connection` method will be called so that a new connection can be created.

All the pieces should be in place now! So let's see if our test finally passes:

```
$ python 04_chat_spies.py TestChatClient
F..
======================================================================
FAIL: test_client_connection (__main__.TestChatClient)
----------------------------------------------------------------------
Traceback (most recent call last):
  File "04_chat_spies.py", line 38, in test_client_connection
    assert connection_spy.broadcast.assert_called_with(("User 1: Hello
World"))
  File "/usr/lib/python3.7/unittest/mock.py", line 873, in
assert_called_with
    raise AssertionError(_error_message()) from cause
AssertionError: Expected call: broadcast('User 1: Hello World')
Actual call: broadcast('Hello World')


----------------------------------------------------------------------
Ran 3 tests in 0.001s

FAILED (failures=1)
```

Unexpectedly, our test failed. The good news is that the connection itself worked. The test was able to put in place the stub, and the spy was used.

The bad news is that our test actually discovered a bug that our previous `TestChatClient.test_send_message` test was unable to spot. In the current implementation of `ChatClient.send_message`, we build the message with the name of the user who wrote it, but we broadcast the one without a name. So none of the other users reading the chat will ever know who wrote that message!

```
class ChatClient:
    ...

    def send_message(self, message):
        sent_message = "{}: {}".format(self.nickname, message)
        self.connection.broadcast(message)
        return sent_message
```

What we want to do here is change the `send_message` method so that the message broadcast is the one with the name of the author, the `sent_message` variable, instead of the `message` one:

```
class ChatClient:
    ...

    def send_message(self, message):
        sent_message = "{}: {}".format(self.nickname, message)
        self.connection.broadcast(sent_message)
        return sent_message
```

Now that we have fixed that bug, our test can finally pass and confirm that our `ChatClient` has the connection in place and properly sends messages through it:

```
$ python 04_chat_spies.py TestChatClient
...
----------------------------------------------------------------------
Ran 3 tests in 0.001s

OK
```

The next step, as usual, is to go back to our acceptance test and ask what's left to do:

```
$ python 04_chat_spies.py TestChatAcceptance
E
======================================================================
ERROR: test_message_exchange (__main__.TestChatAcceptance)
----------------------------------------------------------------------
Traceback (most recent call last):
```

```
  File "04_chat_spies.py", line 10, in test_message_exchange
    user1.send_message("Hello World")
  File "04_chat_spies.py", line 58, in send_message
    self.connection.broadcast(sent_message)
  File "04_chat_spies.py", line 64, in connection
    self._connection = self._get_connection()
  File "04_chat_spies.py", line 74, in _get_connection
    return Connection(("localhost", 9090))
  File "04_chat_spies.py", line 82, in __init__
    self.connect()
  File "/usr/lib/python3.7/multiprocessing/managers.py", line 532, in
connect
    conn = Client(self._address, authkey=self._authkey)
  File "/usr/lib/python3.7/multiprocessing/connection.py", line 492, in
Client
    c = SocketClient(address)
  File "/usr/lib/python3.7/multiprocessing/connection.py", line 619, in
SocketClient
    s.connect(address)
ConnectionRefusedError: [Errno 111] Connection refused


----------------------------------------------------------------
Ran 1 test in 0.001s

FAILED (errors=1)
```

Our acceptance test proves that our client is trying to connect to the server as expected, which is great!

The problem is that, as we already know, there is no such server. So our acceptance test cannot pass as it can't connect to a server and verify that the client is actually able to send and receive messages.

But before moving forward and looking at how to implement our server, let's introduce the concept of mocks, which gather in themselves all the powers of the previously introduced test doubles.

Using mocks

As you've probably noticed, when we use dummy objects, stubs, or spies, we always end up working with the unittest.mock module. That's because mock objects could be seen as dummy objects that provide some stubs mixed with spies.

Mocks are able to be passed around and they usually do nothing, behaving pretty much like **dummy** objects.

If we had a `read_file` function accepting a file object with a `read` method, we could provide a `Mock` instead of a real file; `Mock.read` will just do nothing:

```
>>> def read_file(f):
...     print("READING ALL FILE")
...     return f.read()
...
>>> from unittest.mock import Mock
>>> m = Mock()
>>> read_file(m)
READING ALL FILE
```

If instead of doing nothing, we want to make it act like a **stub**, we can provide a canned response to have `Mock.read` return a predefined string:

```
>>> m.read.return_value = "Hello World"
>>> print(read_file(m))
Hello World
```

If we don't want to just fill in the place of other real objects by replacing them with dummies and stubs, we can also use mocks to track what happened to them, so they are able to behave like a **spy** too:

```
>>> m.read.call_count
2
```

But what makes them mocks is that they can test the behavior of software. Stubs, spies, and dummies are all about state. They provide a state for software consumption when you are injecting a known state into software or for test consumption when you are using a spy to keep the state of calls.

Mocks are usually meant to keep track of behaviors. They usually crash when the software hasn't done what you expected. So they are usually meant to assert that they were used in a specific expected way, which confirms that the software behaved as we wished.

For example, we can check that the `read` method on the `Mock` object was actually called:

```
>>> m.read.assert_called_with()
```

If we wanted to verify that `read_file` was calling `f.read()` with a specific argument, we could have asked the mock to verify that it was used. If the method wasn't called, the assertion would have failed with an `AssertionError`:

```
>>> m.read.assert_called_with("some argument")
Traceback (most recent call last):
  File "<stdin>", line 1, in <module>
  File "/usr/lib/python3.7/unittest/mock.py", line 873, in
assert_called_with
    raise AssertionError(_error_message()) from cause
AssertionError: Expected call: read('some argument')
Actual call: read()
```

If it wasn't called due to a bug or incomplete implementation, the assertion would have detected that and we could have addressed the behavior of `read_file` to make it work as we wanted.

Now that we know about dummies, stubs, spies, and mocks, we know that there are tons of ways to test our software without having to rely on complete and fully functional components. And we know that our test suite has to be fast, easy to debug, and must require minimum dependencies with minimum influence from the external system.

So a real working server would mean having to start a separate server process every time we want to run our tests and would mean slowing down tests because they have to go through a real network connection.

For the next step, instead of implementing a real server, we are going to introduce the concept of **fakes** and try to get a fake server that makes our acceptance test pass.

Replacing dependencies with fakes

Fakes are replacements for real dependencies that are good enough to fake that they are the real deal. Fakes are frequently involved in the goal of simplifying test suite dependencies or improving the performance of a test suite. For example, if your software depends on a third-party weather forecasting API available in the cloud, it's not very convenient to perform a real network connection to the remote API server. The best-case scenario is it will be very slow, and the worst-case scenario is you could get throttled or even banned for doing too many API requests in too short a time, as your test suite could easily reach hundreds or thousands of tests.

The most widespread kind of fakes are usually in-memory databases as they simplify the need to set up and tear down a real database management system for the sole reason of running your tests.

In our case, we don't want to have the need to start a chat server every time we want to run the test suite of our chat application, so we are going to provide a fake server and fake connection that will replace the real networking-based connection.

Now that we have our TestConnection case, which verifies that the connection does what we want, how can we verify that it actually works when there is a server on the other side?

We can look at how the SyncManager server works and provide a fake replacement simple enough to understand the basic protocol and provide the answers. Thankfully, the SyncManager protocol is very simple. It just receives commands with a set of arguments and responds with a tuple, ("RESPONSE_TYPE", RESPONSE), where RESPONSE_TYPE states whether the response is the returned value for that command or an error.

So we can make a FakeServer that provides a FakeServer.send method that will trap the commands that the client is requesting and a FakeServer.recv method that will send back the response to the client:

```python
class FakeServer:
    def __init__(self):
        self.last_command = None
        self.last_args = None
        self.messages = []

    def __call__(self, *args, **kwargs):
        # Make the SyncManager think that a new connection was created.
        return self

    def send(self, data):
        # Track any command that was sent to the server.
        callid, command, args, kwargs = data
        self.last_command = command
        self.last_args = args

    def recv(self, *args, **kwargs):
        # For now we don't support any command, so just error.
        return "#ERROR", ValueError("%s - %r" % (
            self.last_command, self.last_args)
        )

    def close(self):
        pass
```

The very first basic implementation of our fake server is only going to respond to any command with an error, so we can track the commands that the client is trying to send to us.

To test our connection with a server, we are going to add a new
test_exchange_with_server test to the TestConnection test case, which will use the
provided FakeServer to link two connections together:

```
class TestConnection(unittest.TestCase):
    def test_broadcast(self):
        ...

    def test_exchange_with_server(self):
        with unittest.mock.patch(
            "multiprocessing.managers.listener_client",
            new={"pickle": (None, FakeServer())}
        ):
            c1 = Connection(("localhost", 9090))
            c2 = Connection(("localhost", 9090))

            c1.broadcast("connected message")
            assert c2.get_messages()[-1] == "connected message"
```

Our test requires some magic through unittest.mock.patch to replace the standard
implementation of the server/client communication channel in
multiprocessing.managers with our own custom FakeServer. In practice, what we are
doing is replacing the "pickle" based communication channel with our own for the duration
of the test.

Now if we run our test, we should see that our fake server is in place and we should be able
to start tracking which commands are exchanged:

```
$ python 05_chat_fakes.py TestConnection
.E
======================================================================
ERROR: test_exchange_with_server (__main__.TestConnection)
----------------------------------------------------------------------
Traceback (most recent call last):
  File "05_chat_fakes.py", line 56, in test_exchange_with_server
    c1 = Connection(("localhost", 9090))
  File "05_chat_fakes.py", line 100, in __init__
    self.connect()
  File "/usr/lib/python3.7/multiprocessing/managers.py", line 533, in
connect
    dispatch(conn, None, 'dummy')
  File "/usr/lib/python3.7/multiprocessing/managers.py", line 82, in
dispatch
    raise convert_to_error(kind, result)
ValueError: dummy - ()


----------------------------------------------------------------------
```

```
Ran 2 tests in 0.001s

FAILED (errors=1)
```

Our test crashed due to an unrecognized `'dummy'` command (as we currently recognize no commands) but it proved that our fake server is in place and being used by our `Connection` object.

At this point, we can provide support for the `dummy` command (which is just used to establish the connection) and see what happens:

```
class FakeServer:
    ...

    def recv(self, *args, **kwargs):
        if self.last_command == "dummy":
            return "#RETURN", None
        else:
            return "#ERROR", ValueError("%s - %r" % (
                self.last_command, self.last_args)
            )
```

Running again, the `TestConnection` test suite will invoke the next command (after the `"dummy"` one that we just implemented) and thus will complain about the next missing command:

```
$ python 05_chat_fakes.py TestConnection
...
ValueError: create - ('get_messages',)
```

By rerunning our test over and over until it stops crashing, we can spot all the commands that our `FakeServer` has to support in the `FakeServe.recv` method, and one by one, we can implement enough commands to have a fairly complete implementation of our `FakeServer`:

```
class FakeServer:
    ...

    def recv(self, *args, **kwargs):
        if self.last_command == "dummy":
            return "#RETURN", None
        elif self.last_command == "create":
            return "#RETURN", ("fakeid", tuple())
        elif self.last_command == "append":
            self.messages.append(self.last_args[0])
            return "#RETURN", None
        elif self.last_command == "__getitem__":
```

```
            return "#RETURN", self.messages[self.last_args[0]]
        elif self.last_command in ("incref", "decref",
                                   "accept_connection"):
            return "#RETURN", None
        else:
            return "#ERROR", ValueError("%s - %r" % (
                self.last_command, self.last_args)
            )
```

At this point, our `TestConnection` should be able to pass using our fake server to establish the link between the two `Connection` objects:

```
$ python 05_chat_fakes.py TestConnection
..
-------------------------------------------------------------------
Ran 2 tests in 0.001s

OK
```

Our `FakeServer` was able to confirm that the two `Connection` objects are able to talk to each other and see the messages that the other one has broadcast. And we were able to do so without the need to actually start a server instance, listen on the network for the chat connections, and handle that.

While fakes are usually very convenient, the effort required to implement them is frequently pretty high. To be usable, a fake must reproduce a major chunk of the functionalities that the real dependency provided, and as we saw, implementing a fake might involve having to reverse engineer how the piece of software we are trying to replace works.

Luckily, for most widespread needs, you will find fake implementations of SQL servers, MongoDB, S3, and so on, already available as libraries you can install.

While the fake approach worked well, the worst part of our fake usage is probably how we had to patch the `multiprocessing` module to put it in place.

This is a problem caused by the fact that our `Connection` object, being based on `SyncManager`, doesn't provide proper support for dependency injection, which would have allowed us to inject our own communication channel in a proper way instead of having to patch the "pickle" based one.

But before moving on to see how we can handle the injection of dependencies, let's finish our chat application and make our acceptance test pass.

Understanding acceptance tests and doubles

We saw our `Connection` object works with the `FakeServer` but does our acceptance test finally pass now? Not yet. We still have to provide a server there (fake or not) and we still have to finish the implementation of the client.

Acceptance tests are meant to verify that the software really does what we wanted once it's in the hands of our users. For this reason, it's usually a good idea to limit the usage of test doubles in the context of acceptance tests. They should work as much as possible by reproducing the real usage of the software.

While mocks, stubs, dummies, and so on are rarely seen in acceptance tests, it's pretty common to see fakes in that context too. As fakes are supposed to mimic the behavior of the real service they replace, the software should notice no difference. But if you used fakes in your acceptance tests, it's a good idea to introduce a set of system tests that verify the software on the real services it depends on (maybe only executed at release time due to their cost).

In our case, we want our acceptance test to work with a real server, thus we are going to tweak it a little bit to start the server and connect the clients to the newly started server. As our server is implemented on top of a `SyncManager`, like all `SyncManagers` it can be started and stopped by using it as a context manager in a `with` statement.

When we enter the `with new_chat_server()` context, the server will be started, and once we exit it, the server will be stopped:

```
class TestChatAcceptance(unittest.TestCase):
    def test_message_exchange(self):
        with new_chat_server():
            user1 = ChatClient("John Doe")
            user2 = ChatClient("Harry Potter")

            user1.send_message("Hello World")
            messages = user2.fetch_messages()
            assert messages == ["John Doe: Hello World"]
```

Obviously, running the test will fail because we have not yet made the `new_chat_server` function that is supposed to return the server in use by the test.

Our server will be just a `SyncManager` subclass that provides the list of messages (through the `_srv_get_messages` function) so that the clients can access them:

```
_messages = []
def _srv_get_messages():
    return _messages
class _ChatServerManager(SyncManager):
    pass
_ChatServerManager.register("get_messages",
                            callable=_srv_get_messages,
                            proxytype=ListProxy)

def new_chat_server():
    return _ChatServerManager(("", 9090), authkey=b'mychatsecret')
```

Now that we've created our `new_chat_server`, which can be used to start the server, our next step is, as usual, to verify that our tests do pass to see what's the next step:

```
$ python 06_acceptance_tests.py TestChatAcceptance
E
======================================================================
ERROR: test_message_exchange (__main__.TestChatAcceptance)
----------------------------------------------------------------------
Traceback (most recent call last):
  File "06_dependency_injection.py", line 12, in test_message_exchange
    messages = user2.fetch_messages()
AttributeError: 'ChatClient' object has no attribute 'fetch_messages'

----------------------------------------------------------------------
Ran 1 test in 0.011s

FAILED (errors=1)
```

In this case, the test doesn't yet pass because we forgot to implement the last piece of our client: the part related to fetching the messages. So let's add that new `fetch_messages` method to our client and see if things work as we want.

As usual, we should start with a test for the `ChatClient.send_message` unit, so that we can verify that our implementation does what we expect:

```
class TestChatClient(unittest.TestCase):
    ...

    def test_client_fetch_messages(self):
        client = ChatClient("User 1")
        client.connection = unittest.mock.Mock()
        client.connection.get_messages.return_value = ["message1",
```

```
                                                           "message2"]
    starting_messages = client.fetch_messages()
    client.connection.get_messages().append("message3")
    new_messages = client.fetch_messages()

    assert starting_messages == ["message1", "message2"]
    assert new_messages == ["message3"]
```

As our `ChatClient.fetch_messages` method doesn't yet exist, our test unit will immediately fail:

```
$ python 06_acceptance_tests.py TestChatClient
.E..
=======================================================================
ERROR: test_client_fetch_messages (__main__.TestChatClient)
-----------------------------------------------------------------------
Traceback (most recent call last):
  File "06_dependency_injection.py", line 46, in test_client_fetch_messages
    starting_messages = client.fetch_messages()
AttributeError: 'ChatClient' object has no attribute 'fetch_messages'


-----------------------------------------------------------------------
Ran 4 tests in 0.001s

FAILED (errors=1)
```

So, what we can do is go back to `ChatClient` and implement the `fetch_messages` method in a way that satisfies our test:

```python
class ChatClient:
    def __init__(self, nickname):
        self.nickname = nickname
        self._connection = None
        self._last_msg_idx = 0

    def send_message(self, message):
        sent_message = "{}: {}".format(self.nickname, message)
        self.connection.broadcast(sent_message)
        return sent_message

    def fetch_messages(self):
        messages = list(self.connection.get_messages())
        new_messages = messages[self._last_msg_idx:]
        self._last_msg_idx = len(messages)
        return new_messages
```

The new `ChatClient.fetch_messages` method will fetch the messages stored by the server and will return any new ones since the last time they were checked.

If our implementation is correct, running the test again will make it pass and will confirm that our method does what we wanted it to do:

```
$ python 06_acceptance_tests.py TestChatClient
....
---------------------------------------------------------------------
Ran 4 tests in 0.001s

OK
```

Also, as this was our last missing piece, the acceptance test should now pass, confirming that our chat application does work as we wanted:

```
$ python 06_acceptance_tests.py TestChatAcceptance
.
---------------------------------------------------------------------
Ran 1 test in 0.016s

OK
```

Hurray! We can finally declare victory. Our application works with the real client and real server. They are able to connect and talk to each other, which proves we wrote the software we wanted to write.

But our `ChatClient` tests have fairly complex code that has to rely on `mock.patch` to replace pieces of it and we even had to implement a property setter for the connection for the sole purpose of making it possible to replace it with a testing double.

Even though we achieved our goal, there should be a better way to enable test doubles in code than spreading `mock.patch` everywhere.

Replacing components of a system on demand is what **dependency injection** was made for, so let's see if it can help us to switch between using fakes and real services in our test suite.

Managing dependencies with dependency injection

Our `ChatClient` machinery to connect to a server is rather more complex than necessary. The `ChatClient.get_connection` and `ChatClient.connection` property setters are there mostly to allow us to easily replace with mocks the connections that our client sets up.

This is because `ChatClient` has a dependency, a dependency on the `Connection` object, and it tries to satisfy that dependency all by itself. It's like when you are hungry... You depend on food to solve your need, so you go to the fridge, take some ingredients, turn on the oven, and cook a meal yourself. Then you can eat. Or... you can call a restaurant and order a meal.

Dependency injection gives you a way to take the restaurant path. If your `ChatClient` needs a connection, instead of trying to get a connection itself, it can ask for a connection and someone else will take care of providing it.

In most dependency injection systems, there is an **injector** that will take care of getting the right object and providing it to the client. The client typically doesn't even have to know about the injector. This usually involves fairly advanced frameworks that provide a services registry and allow clients to register for those services, but there is a very simple form of dependency injection that works very well and can be immediately achieved without any external dependency or framework: **construction injection**.

Construction injection means that the service your code depends on is provided as a parameter when building the class that depends on it.

In our case, we could easily refactor the `ChatClient` to accept a `connection_provider` argument, which would allow us to simplify our `ChatClient` implementation and get rid of entire parts of it:

```python
class ChatClient:
    def __init__(self, nickname, connection_provider=Connection):
        self.nickname = nickname
        self._connection = None
        self._connection_provider = connection_provider
        self._last_msg_idx = 0

    def send_message(self, message):
        sent_message = "{}: {}".format(self.nickname, message)
        self.connection.broadcast(sent_message)
        return sent_message
```

```
def fetch_messages(self):
    messages = list(self.connection.get_messages())
    new_messages = messages[self._last_msg_idx:]
    self._last_msg_idx = len(messages)
    return new_messages

@property
def connection(self):
    if self._connection is None:
        self._connection = self._connection_provider(("localhost",
                                                       9090))

    return self._connection
```

We got rid of `ChatClient.get_connection` and we got rid of the `connection` `@property.setter` but we haven't lost a single functionality, nor have we added any additional complexity. In most cases, the `ChatClient` can be used exactly like before and it will take care of using the right `Connection` by default.

But for the cases where we want to do something different, we can inject other kinds of connections.

For example, in our `TestChatClient.test_client_connection` test, we can remove a fairly hard-to-read `mock.patch` that was in place to set up a spy:

```
class TestChatClient(unittest.TestCase):
    def test_client_connection(self):
        client = ChatClient("User 1")

        connection_spy = unittest.mock.MagicMock()
        with unittest.mock.patch.object
          (client, "_get_connection", return_value=connection_spy):
            client.send_message("Hello World")

        connection_spy.broadcast.assert_called_with(("User 1:
                                                      Hello World"))
```

Instead of having to patch the implementation of ChatClient, we can just provide the spy to the ChatClient and have it use it:

```python
def test_client_connection(self):
    connection_spy = unittest.mock.MagicMock()

    client = ChatClient("User 1", connection_provider=lambda *args:
                        connection_spy)
    client.send_message("Hello World")

    connection_spy.broadcast.assert_called_with(("User 1:
                                                Hello World"))
```

The code is far easier to follow and understand and doesn't rely on magic such as patching objects at runtime.

In fact, our whole TestChatClient can be made simpler by using dependency injection instead of patching:

```python
class TestChatClient(unittest.TestCase):
    def test_nickname(self):
        client = ChatClient("User 1")

        assert client.nickname == "User 1"

    def test_send_message(self):
        client = ChatClient("User 1",
                            connection_provider=unittest.mock.Mock())

        sent_message = client.send_message("Hello World")
        assert sent_message == "User 1: Hello World"

    def test_client_connection(self):
        connection_spy = unittest.mock.MagicMock()

        client = ChatClient("User 1", connection_provider=lambda *args:
                            connection_spy)
        client.send_message("Hello World")

        connection_spy.broadcast.assert_called_with(("User 1: Hello
                                                    World"))

    def test_client_fetch_messages(self):
        connection = unittest.mock.Mock()
        connection.get_messages.return_value = ["message1", "message2"]

        client = ChatClient("User 1", connection_provider=lambda *args:
                            connection)
```

```
starting_messages = client.fetch_messages()
client.connection.get_messages().append("message3")
new_messages = client.fetch_messages()

assert starting_messages == ["message1", "message2"]
assert new_messages == ["message3"]
```

In all cases where we had fairly hard-to-read uses of `mock.patch`, we have now replaced them with an explicitly provided `connection_provider` when the `ChatClient` is created.

So dependency injection can make your life easier when testing, but actually also makes your implementation far more flexible.

Suppose that we want to have our chat app working on something different than `SyncManagers`; now it's a matter of just passing a different kind of `connection_provider` to our clients.

Whenever your classes depend on other objects that they are going to build themselves, it's usually a good idea to question whether that's a place for dependency injection and whether those services could be injected from outside instead of being built within the class itself.

Using dependency injection frameworks

In Python, there are many frameworks for dependency injection, and it's an easy enough technique to implement yourself that you will find various variations of it in many frameworks. What dependency injection frameworks will do for you is wire the objects together.

In our previous dependency injection paragraph, we explicitly provided the dependencies every time we wanted to create a new object (apart from the default dependency, which was provided for us, being the default argument). A dependency injection framework would instead automatically detect for us that `ChatClient` needs a `Connection` and it would give the connection to the `ChatClient`.

One of the easiest-to-use dependency injection frameworks for Python is `Pinject` from Google. It comes from the great experience Google teams have with dependency injection frameworks, which is clear if you look at some of their most famous frameworks, such as Angular.

`Pinject` manages dependencies in a very simple and easy to understand way, based on initializer argument names and class names.

Suppose that, like before, we had our two `ChatClient` and `Connection` classes... but in this case, our `ChatClient` is just going to print which `Connection` it's going to use, as our sole purpose is to showcase how `Pinject` can handle dependency injection for us:

```
class ChatClient:
    def __init__(self, connection):
        print(self, "GOT", connection)

class Connection:
    pass
```

Then we can use `pinject` to create a graph of the dependencies of our objects:

```
import pinject
injector = pinject.new_object_graph()
```

Once `pinject` is aware of the dependencies of our objects (which by default are built by scanning all classes in all imported modules; you can also pass your classes explicitly through a `classes=` argument), we can ask `pinject` to give us an instance for any class it's aware of, resolving all class dependencies for us:

```
>>> cli = injector.provide(ChatClient)
<ChatClient object at 0x7fad51469610> GOT <Connection object at
0x7fad51469bd0>
```

What happened is that pinject detected that a `Connection` class existed and when we requested a `ChatClient`, it saw that it depended on a `Connection` argument. At that point, pinject automatically made a connection for us and provided it to the client.

What if we wanted to provide a fake `Connection` object for our tests? Pinject supports providing custom binding specifications, so telling it explicitly which class solves a specific dependency.

So if we had a `FakeConnection` object, we could create a `pinject.BindingSpec` to tell pinject that to satisfy the `"connection"` dependency, it has to use the fake one:

```
class FakeConnection:
    pass

class FakedBindingSpec(pinject.BindingSpec):
    def provide_connection(self):
        return FakeConnection()
```

```
faked_injector = pinject.new_object_graph(binding_specs=[
    FakedBindingSpec()
])
```

At this point, if we tried to create a ChatClient through the faked_injector, we would get back a ChatClient that uses a fake connection:

```
>>> cli = faked_injector.provide(ChatClient)
<ChatClient object at 0x7fad513ce350> GOT <FakeConnection object at
0x7fad513d6f90>
```

It must be noted that, by default, Pinjector remembers the instances it made, so if we requested a new ChatClient, it would get the same exact connection object. That is frequently convenient when you are building a full piece of software and you want to replace whole components. If you wanted to replace your data abstraction layer to use a fake database, you would probably want to get the same data abstraction layer from everywhere so that all components see the same data.

This means that creating a new ChatClient will give us a different ChatClient but with the same underlying Connection:

```
>>> cli = faked_injector.provide(ChatClient)
<ChatClient object at 0x7f9878aeb810> GOT <Connection object at
0x7f9878a58f50>
>>> cli2 = faked_injector.provide(ChatClient)
<ChatClient object at 0x7f9878a55fd0> GOT <Connection object at
0x7f9878a58f50>
```

In the case of our clients, we probably want each of them to have a different connection to the server. To do so, we can use the BindingSpec and tell pinject that our returned dependency is a prototype and not a singleton. This way, pinject won't cache the provided dependency and will always return a new one:

```
class PrototypeBindingSpec(pinject.BindingSpec):
    @pinject.provides(in_scope=pinject.PROTOTYPE)
    def provide_connection(self):
        return Connection()

proto_injector = pinject.new_object_graph(binding_specs=[
    PrototypeBindingSpec()
])
```

If we were to make a `ChatClient` with the `proto_inject`, we would now see that each client has its own `Connection` object:

```
>>> cli = proto_injector.provide(ChatClient)
<ChatClient object at 0x7fadab060e50> GOT <Connection object at
0x7fadab013910>
>>> cli2 = proto_injector.provide(ChatClient)
<ChatClient object at 0x7fadab060f10> GOT <Connection object at
0x7fadab013850>
```

So, dependency injection frameworks can solve many needs for you. Whether you need to use one or not depends mostly on how complex the network of dependencies in your software is, but having one around can usually give you a quick way to break dependencies between your components when you need to.

Summary

Dependencies between the components that you have to test can make your life hard as a developer. To test anything more complex than a simple utility function, you might end up having to cope with tens of dependencies and their state.

This is why the idea of being able to provide doubles for testing in place of the real components was quickly born once the idea of automated tests became reality. Being able to replace the components the unit you are testing depends on with fakes, dummies, stubs, and mocks can make your life a lot easier and keep your test suite fast and easy to maintain.

The fact that any software is, in reality, a complex network of dependencies is the reason why many people advocate that integration tests are the most realistic and reliable form of testing, but managing that complex network can be hard and that's where dependency injection and dependency injection frameworks can make your life far easier.

Now that we know how to write automatic test suites and we know how to use test doubles to verify our components in isolation and spy their state and behavior, we have all the tools that we need to dive into test-driven development in the next chapter and see how to write software in the TDD way.

3
Test-Driven Development while Creating a TODO List

No programmer ever releases a software without having tested it – even for the most basic proof of concept and rough hack, the developer will run it once to see that it at least starts and resembles what they had in mind.

But to **test**, as a verb, usually ends up meaning clicking buttons here and there to get a vague sense of confidence that the software does what we intended. This is different from **test** as a noun, which means a set of written-out checks that our software must pass to confirm it does what we wanted.

Apart from being more reliable, written-out checks force us to think about what the code must do. They force us to get into the details and think beforehand about what we want to build. Otherwise, we would just jump to building without thinking about what we are building. And trying to ensure that what gets built is, in every single detail, the right thing through a written specification is quickly going to turn into writing the software itself, just in plain English.

The problem is that the more hurried, stressed, and overwhelmed developers are, the less they test. Tests are the first thing that get skipped when things go wrong, and by doing so things suddenly get even worse, as tests are what avoid errors and failures, and more errors and failures mean more stress and rushing through the code to fix them, making the whole process a loop that gets worse and worse.

Test-Driven Development (TDD) tries to solve this problem by engendering a set of practices where tests become a fundamental step of your daily routine. To write more code you must write tests, and as you get used to TDD and it becomes natural, you will quickly notice that it gets hard to even think about how to get started if not by writing a test.

That's why in this chapter, we will cover how TDD can fit into the software development routine and how to leverage it to keep problems under control at times of high stress.

In this chapter, we will cover the following topics:

- Starting projects with TDD
- Building applications the TDD way
- Preventing regressions

Technical requirements

A working Python interpreter should be all that is needed to work through the exercises in this chapter.

The examples have been written using Python 3.7, but should work on most modern Python versions.

You can find the code files used in this chapter on GitHub at `https://github.com/PacktPublishing/Crafting-Test-Driven-Software-with-Python/tree/main/Chapter03`

Starting projects with TDD

We already know that tests are meant to verify that our software adheres to the desired behavior. To do so means that our tests must express what that desired behavior is. They must explicitly state, *"If I do this, I expect that to happen."*

For the innermost components, what happens is probably an implementation detail: *"If I commit my unit of work, data is written to the database."* But the more we move to the outer parts of our architecture, those that connect our software to the outside world, the more these tests become expressions of business needs. The more we move from solitary units, to sociable units, to integration and acceptance tests, the more the "desired behavior" becomes the one that has a business value.

If we work with a test-driven approach, our first step before writing implementation code is obviously to write a test that helps us understand what we want to build (if we are just starting with our whole project, what we want to build is the software itself). This means that our very first test is the one that is going to make clear what's valuable. Why are we even writing the software in the first place?

So let's see how a test-driven approach can benefit us during the software design phase itself. Suppose we want to start a TODO list kind of product.

So let's start writing an acceptance test that will help us express explicitly what we want our app to do.

Let's create a new `todo` directory where we are going to put the `todo/src` subdirectory with our source code, and the `todo/tests` directory with our tests:

```
$ tree
.
├── src
└── tests
```

At this point, we can start by making a `todo/tests/__init__.py` file and a `todo/tests/test_acceptance.py` module for our overall application acceptance test. The `test_acceptance.py` file is going to contain our test itself:

```
import unittests


class TestTODOAcceptance(unittest.TestCase):
    def test_main(self):
        raise NotImplementedError()
```

We want our interactive shell application to accept commands and print outputs. So the first thing we want the app to do is to write the output and receive commands from an input:

```
class TestTODOAcceptance(unittest.TestCase):
    def test_main(self):
        app = TODOApp(io=(self.fake_input, self.fake_output))
```

We don't yet know what our `fake_input` and `fake_output` will be, but we will figure that out as we reduce uncertainty about how the app should behave.

Then we said we want it to be an interactive shell, so it should be sitting there accepting commands until we tell it to quit. To make that happen we probably want to have the main loop for our **Read-Eval-Print Loop** (**REPL**) and we want the app to be running in the background during our test so we can send commands to it and fetch the responses:

```
import unittest
import threading


class TestTODOAcceptance(unittest.TestCase):
    def test_main(self):
        app = TODOApp(io=(self.fake_input, self.fake_output))
```

```
app_thread = threading.Thread(target=app.run, daemon=True)
app_thread.start()
```

But we don't want our app to be stuck there forever until the user kills it abruptly due to the frustration of being unable to exit it, and we surely don't want our test to be stuck there forever either. So we want our app to support a `quit` command and ensure it exits when it receives it:

```python
import unittest
import threading


class TestTODOAcceptance(unittest.TestCase):
    def test_main(self):
        app = TODOApp(io=(self.fake_input, self.fake_output))

        app_thread = threading.Thread(target=app.run, daemon=True)
        app_thread.start()

        # ...

        self.send_input("quit")
        app_thread.join(timeout=1)
        self.assertEqual(self.get_output(), "bye!\n")
```

Great, now we know we want our app to sit there, accept commands, and exit on a quit request. But how are we going to tell the user that we are accepting commands? We likely want a prompt, so let's verify that by presenting a welcome screen with the list of the TODO items (none at the beginning) and a "> " prompt:

```python
import unittest
import threading


class TestTODOAcceptance(unittest.TestCase):
    def test_main(self):
        app = TODOApp(io=(self.fake_input, self.fake_output))

        app_thread = threading.Thread(target=app.run, daemon=True)
        app_thread.start()

        welcome = self.get_output()
        self.assertEqual(welcome, (
            "TODOs:\n"
            "\n"
            "\n"
            "> "
```

```
        ))

        self.send_input("quit")
        app_thread.join(timeout=1)
        self.assertEqual(self.get_output(), "bye!\n")
```

Very well, we've already provided answers to tons of questions about how our app should behave. We decided it's driven by commands and those commands can be provided through a prompt on the same screen that displays the list of our items.

What primary commands do we want to provide? Surely we at least want to be able to add new items and delete them? So let's test that we can execute those commands:

```
import unittest
import threading


class TestTODOAcceptance(unittest.TestCase):
    def test_main(self):
        app = TODOApp(io=(self.fake_input, self.fake_output))

        app_thread = threading.Thread(target=app.run, daemon=True)
        app_thread.start()

        welcome = self.get_output()
        self.assertEqual(welcome, (
            "TODOs:\n"
            "\n"
            "\n"
            "> "
        ))

        self.send_input("add buy milk")
        welcome = self.get_output()
        self.assertEqual(welcome, (
            "TODOs:\n"
            "1. buy milk\n"
            "\n"
            "> "
        ))

        self.send_input("add buy eggs")
        welcome = self.get_output()
        self.assertEqual(welcome, (
            "TODOs:\n"
            "1. buy milk\n"
            "2. buy eggs\n"
            "\n"
```

```
            "> "
    ))

    self.send_input("del 1")
    welcome = self.get_output()
    self.assertEqual(welcome, (
        "TODOs:\n"
        "1. buy eggs\n"
        "\n"
        "> "
    ))

    self.send_input("quit")
    app_thread.join(timeout=1)
    self.assertEqual(self.get_output(), "bye!\n")
```

OK, we added a block where we add a note to "buy milk", one with a note to "buy eggs", and a third where we delete the "buy milk" entry. Our acceptance test is now fairly complete! It adds multiple todos and it removes them. We've defined everything we want our app to do and we can now move forward to finally trying to satisfy our needs!

The test itself is going to do the next step; we simply have to run it:

```
$ python -m unittest discover
E
======================================================================
ERROR: test_main (tests.test_acceptance.TestTODOAcceptance)
----------------------------------------------------------------------
Traceback (most recent call last):
  File "/testingbook/03_specifications/01_todo/tests/test_acceptance.py",
line 20, in test_main
    app = TODOApp(io=(self.fake_input, self.fake_output))
NameError: name 'TODOApp' is not defined


----------------------------------------------------------------------
Ran 1 test in 0.000s

FAILED (errors=1)
```

Right, we now need to make the app itself as it doesn't even exist yet as a concept.

For the sake of focusing this section on the business value of our application and thus on the user-facing tests, we are going to diverge a bit from the correct approach and we are going to have a single test for the whole app.

So don't be surprised if we jump here from an acceptance test directly to the implementation of the app itself. It's only for the sake of reducing the cognitive load of the reader. In the real world, we would be writing unit tests to drive the design of our code, as designing the app and designing its implementation are two very different things. But here we wanted to make clear how writing tests forces us to think about the app itself, and thus we are going to make the code design happen behind the scenes.

In the *Building applications the TDD way* section, we are going to see how to mix what we learned here about acceptance tests with the more classical TDD approach regarding the design of the code itself.

So let's create a new `todo` Python package inside our `src` directory. We are going to have a `src/todo/__init__.py` file and a `src/todo/app.py` module for the application implementation itself:

```
$ tree
.
├── src
│   ├── todo
│   │   ├── app.py
│   │   ├── __init__.py
│   └── tests
│       ├── __init__.py
│       └── test_acceptance.py
```

Our `TODOApp` can reside in `src/todo/app.py` for now, just as an empty class:

```
class TODOApp:
    pass
```

Is this enough to be able to use our app from the tests? Not yet, because the `todo` package is not available for our tests. So before moving forward, we want to add a `src/setup.py` file to make a distribution for our `todo` package. Our minimal `setup.py` file is just going to tell the Python installer that *"The application is named todo and it contains a todo package that has to be installed"*:

```
from setuptools import setup

setup(name='todo', packages=['todo'])
```

Then the final layout of our project directory should look pretty much like this:

```
$ tree
.
├── src
│   ├── setup.py
│   ├── todo
│   │   ├── app.py
│   │   └── __init__.py
│   └── tests
        ├── __init__.py
        └── test_acceptance.py
```

At this point, we can install our application in development mode with `pip install -e`:

```
$ pip install -e src/
Obtaining file://testingbook/03_specifications/01_todo/src
Installing collected packages: todo
  Running setup.py develop for todo
Successfully installed todo
```

This allows us to edit our `tests/test_acceptance.py` file to import the application class itself and solve the previous `NameError` error:

```
import unittest

from todo.app import TODOApp


class TestTODOAcceptance(unittest.TestCase):
    def test_main(self):
        ...
```

We already know that our `TODOApp` does nothing, so it surely won't make our test pass, but let's see what our test suggests for the next required step that involves rerunning our test suite:

```
$ python -m unittest discover
======================================================================
ERROR: test_main (tests.test_acceptance.TestTODOAcceptance)
...
    app = TODOApp(io=(self.fake_input, self.fake_output))
AttributeError: 'TestTODOAcceptance' object has no attribute 'fake_input'
```

Given that we've now installed the `todo` package, the app imports fine, but the test has no `fake_input` and `fake_output` to provide. So those are going to be our next areas of attention.

As we want to ship input and outputs back and forth between the test and the app, wait for the outputs to be available, and use something that works across threads, a well-fitting solution might be to use a queue. During the application execution, our output function will probably be the `print` function and our input will be the Python `input` function, so let's set up something that allows us to simulate those.

In our test case setup, we are going to create the **Input/Output (I/O)** queues and create a `self.fake_input` object that simulates the behavior of input and a `self.fake_output` object that simulates the behavior of `print`. Also for convenience, we are going to add the `self.get_output` and `self.send_input` methods so that our test can send and receive text from the app:

```python
import unittest
import threading
import queue

from todo.app import TODOApp


class TestTODOAcceptance(unittest.TestCase):
    def setUp(self):
        self.inputs = queue.Queue()
        self.outputs = queue.Queue()

        self.fake_output = lambda txt: self.outputs.put(txt)
        self.fake_input = lambda: self.inputs.get()

        self.get_output = lambda: self.outputs.get(timeout=1)
        self.send_input = lambda cmd: self.inputs.put(cmd)

    def test_main(self):
        app = TODOApp(io=(self.fake_input, self.fake_output))
        ...
```

OK, we should have in place our I/O infrastructure for the tests. Will our test move forward? Let's see:

```
$ python -m unittest discover
================================================================
ERROR: test_main (tests.test_acceptance.TestTODOAcceptance)
...
    app = TODOApp(io=(self.fake_input, self.fake_output))
TypeError: TODOApp() takes no arguments
```

OK, not as much as hoped. It did move forward, but we crashed on the same exact line of code because our `TODOApp` doesn't yet have any concept of I/O.

So let's make our `TODOApp` aware of its input and output. By default, we are going to provide the built-in Python `input` and `print` commands (without the trailing newline), but our test will replace those with its own `fake_input` and `fake_output`:

```
import functools

class TODOApp:
    def __init__(self, io=(input, functools.partial(print, end=""))):
        self._in, self._out = io
```

OK, we now have a `TODOApp._in` callable we can use to ask for inputs, and a `TODOApp._out` callable we can use to write outputs. What's the next step?

```
$ python -m unittest discover
========================================================================
ERROR: test_main (tests.test_acceptance.TestTODOAcceptance)
...
    app_thread = threading.Thread(target=app.run)
AttributeError: 'TODOApp' object has no attribute 'run'
```

Right, the REPL! Our app needs to leverage those I/O functions to actually show the output and ask for inputs. So we are going to add a `TODOApp.run` function that runs our REPL, providing the prompt and accepting commands until we quit:

```
import functools


class TODOApp:
    def __init__(self, io=(input, functools.partial(print, end=""))):
        self._in, self._out = io
        self._quit = False

    def run(self):
        self._quit = False
        while not self._quit:
            self._out(self.prompt(""))
            command = self._in()
        self._out("bye!\n")

    def prompt(self, output):
        return """TODOs:
{}

> """.format(output)
```

For now, our interactive shell doesn't do much – it shows the prompt and does nothing with the commands we send.

If we run our acceptance test again, we are going to clearly see that our app did receive the add command to add the `buy milk` entry, but it didn't execute it and so the entry isn't there:

```
$ python -m unittest discover
=====================================================================
...
AssertionError: 'TODOs:\n\n\n> ' != 'TODOs:\n1. buy milk\n\n> '
  TODOs:
-
+ 1. buy milk
```

Our next step is adding the command dispatching and execution functionality so that the REPL not only receives those commands, but also executes them:

```python
import functools


class TODOApp:
    def __init__(self, io=(input, functools.partial(print, end=""))):
        self._in, self._out = io
        self._quit = False

    def run(self):
        self._quit = False
        while not self._quit:
            self._out(self.prompt(""))
            command = self._in()
            self._dispatch(command)
        self._out("bye!\n")

    def prompt(self, output):
        return """TODOs:
{}

> """.format(output)

    def _dispatch(self, cmd):
        cmd, *args = cmd.split(" ", 1)
        executor = getattr(self, "cmd_{}".format(cmd), None)
        if executor is None:
            self._out("Invalid command: {}\n".format(cmd))
            return
        executor(*args)
```

The `TODOApp.run` method is in charge of calling `TODOApp._dispatch` to serve commands, and each command will be served by running a `TODOApp.cmd_COMMANDNAME` method that we will implement for each command.

If we rerun our test, we are going to get complaints about invalid commands being sent to the application:

```
$ python -m unittest discover
======================================================================
FAIL: test_main (tests.test_acceptance.TestTODOAcceptance)
...
AssertionError: 'Invalid command: add\n' != 'TODOs:\n1. buy milk\n\n> '
- Invalid command: add
+ TODOs:
+ 1. buy milk
+
+ >
```

This is pretty much expected because we have not yet implemented any commands.

So let's provide our `add` command, which is simply going to get the entry to add and insert the todo item into the list of our TODO entries:

```python
class TODOApp:
    def __init__(self, io=(input, functools.partial(print, end=""))):
        self._in, self._out = io
        self._quit = False
        self._entries = []

    ...

    def cmd_add(self, what):
        self._entries.append(what)
```

Rerunning our acceptance test will confirm that the `Invalid command` message went away, and thus we can now handle the command, but we still don't print back the list of todo items. So even if the todo entry was added to our todo list, it's not displayed back to us:

```
$ python -m unittest discover
========================================================================
FAIL: test_main (tests.test_acceptance.TestTODOAcceptance)
...
AssertionError: 'TODOs:\n\n\n> ' != 'TODOs:\n1. buy milk\n\n> '
  TODOs:
-
+ 1. buy milk
  >
```

Instead of showing an empty prompt, like the current `self.prompt("")` call is doing, we want to actually show the list of our TODO items. So we are going to add an `items_list` method to our `TODOApp` that returns the content we want to display in the prompt through `self.prompt(self.items_list())` during the REPL loop within `TODOApp.run`:

```
class TODOApp:
    def __init__(self, io=(input, functools.partial(print, end=""))):
        self._in, self._out = io
        self._quit = False
        self._entries = []

    def run(self):
        self._quit = False
        while not self._quit:
            self._out(self.prompt(self.items_list()))
            command = self._in()
            self._dispatch(command)
        self._out("bye!\n")

    def items_list(self):
        enumerated_items = enumerate(self._entries, start=1)
        return "\n".join(
            "{}. {}".format(idx, entry) for idx, entry in enumerated_items
        )

    ...
```

Our application will now be able to finally serve its first complete cycle, receiving the `add` command and showing us the list of items with the newly added entry.

If we rerun our test, we no longer get stuck on the same issue of having an empty list of todo items, but we are going to get complaints about the fact that the `del` command is not yet implemented:

```
$ python -m unittest discover
===========================================================================
FAIL: test_main (tests.test_acceptance.TestTODOAcceptance)
...
AssertionError: 'Invalid command: del\n' != 'TODOs:\n1. buy eggs\n\n> '
- Invalid command: del
+ TODOs:
+ 1. buy eggs
+
+ >
```

So let's implement the remaining two commands, `del` and `quit`, and check whether our app is complete:

```python
class TODOApp:
    ...

    def cmd_quit(self, *_):
        self._quit = True

    def cmd_add(self, what):
        self._entries.append(what)

    def cmd_del(self, idx):
        idx = int(idx) - 1
        if idx < 0 or idx >= len(self._entries):
            self._out("Invalid index\n")
            return

        self._entries.pop(idx)
    ...
```

The `cmd_del` function just checks whether a valid index to be removed was provided, and then removes it from the list of todo entries. The `cmd_quit` command just sets a flag that will make our REPL exit when it finds it on the next loop cycle.

Now that the functionality to add todo items, remove them, and quit the app has been implemented, our test will finally succeed and confirm our application matches our requirements:

```
$ python -m unittest discover
.
---------------------------------------------------------------
```

```
Ran 1 test in 0.001s
```

```
OK
```

So far, we made an entire application without launching it even once. We had the whole implementation driven by our acceptance test. Will the app really work and do what we wanted? Did acceptance tests really help us design the application behavior?

To check whether the experience is the one we expected, let's make our application runnable. This can be done by adding a __main__.py file to our todo package within src/todo. The updated result of our project layout should thus be as follows:

```
$ tree
.
├── src
│   ├── setup.py
│   └── todo
│       ├── app.py
│       ├── __init__.py
│       └── __main__.py
└── tests
    ├── __init__.py
    └── test_acceptance.py

3 directories, 6 files
```

And the content of src/todo/__main__.py will be very simple — it will just create our TODOApp and will enter the main loop:

```
from .app import TODOApp

TODOApp().run()
```

Our app can now be started with the python -m todo command. Let's see whether the behavior is actually what we imagined and our test-driven design approach really leads to the app we expected:

```
$ python -m todo
TODOs:

> add buy some milk
TODOs:
1. buy some milk

> add buy water
TODOs:
1. buy some milk
```

```
2. buy water

> add send happy birthday message
TODOs:
1. buy some milk
2. buy water
3. send happy birthday message

> del 1
TODOs:
1. buy water
2. send happy birthday message

> del 1
TODOs:
1. send happy birthday message

> quit
bye!
```

Definitely, the app behaves as we expected! We were welcomed by a prompt with an empty list of todo items and as we added and removed them, our prompt updated with the new state of our todo list. The app delivered exactly the experience we described in our test and supports all the features we wanted, working flawlessly on the first run.

This approach of driving the whole software design and development process from business-oriented acceptance tests usually comes under the umbrella of **Acceptance Test-Driven Development (ATDD)**.

We saw how tests not only verify the correctness of the software but at the outer layers, can also explain what the primary software behaviors are and what the software's business value is.

This means that tests can tell a story – if I read them, I'm going to know exactly how the software behaves in that context. If the software has a good enough test coverage and I read all the tests, then I'm going to know how the software works as a whole. Thus tests can be used to express the software specification itself in a reliable and testable manner, which is a concept frequently referred to as **Specification by Example.**

We are going to get into more details about this concept in Chapter 7, *Fitness Function with a Contact Book Application*, but for now, let's focus on how to attach this concept of designing the software through tests to the concept of designing its implementation through tests.

Building applications the TDD way

In the previous section, we saw how to use tests to design our application itself, exposing clear goals and forcing us to think about how the application should behave.

Once we start thinking a bit about what a test is actually doing, it slowly becomes clear why that works well: the tests are going to interact with the system under test. The way they are going to interact with the system they have to test is usually through the interface that the system exposes.

This means that the capabilities we are going to expose to any black-box test are the same capabilities that we are going to expose to any other user of the system under test.

If the system under test is the whole application, as in the case of the previous section, then it means that to write the test we will be forced to reason about the capabilities and the interface we are going to expose to our users themselves. In practice, having to write a test for that layer forces us to make clear the UI and UX of our application.

If the system under test is instead a component of the whole application, the user of that component will be another software component; another piece of code calling the first one. This means that to write the test, we will be forced to define the API that our component has to expose, and thus design the implementation of the component itself.

Thus embracing TDD helps us design code with well-thought-out APIs that the rest of the system can depend on, but writing tests beforehand is not the sum of all TDD practices. There are two primary rules that are part of the TDD practice: the first is obviously to write failing tests before you write the code, but the second is that once your tests pass, you should refactor to remove duplication.

This means that it not only forces us to think of the public interfaces that our objects and subsystems are going to expose beforehand, but it also forces us to keep our internals in shape through continuous refactoring.

The TODO list application we made does everything we wanted, but it lacks a fairly major feature before it can become a valuable application we can use for real: it doesn't persist our todo items. If we close the application and restart it, we are going to lose all our items.

We definitely want our TODO app to save and reload our todo items, so we are going to work on a new feature to enable that behavior.

As usual, we are going to start with a very high-level acceptance test that shows what we want the experience for the user to be. Our new `test_persistence` test is going to start a new todo app with an empty database, save an item, quit the app, and restart it again on the same database to check that the items are still there:

```
...
import tempfile


class TestTODOAcceptance(unittest.TestCase):
    ...

    def test_persistence(self):
        with tempfile.TemporaryDirectory() as tmpdirname:
            app_thread = threading.Thread(
                target=TODOApp(
                    io=(self.fake_input, self.fake_output),
                    dbpath=tmpdirname
                ).run,
                daemon=True
            )
            app_thread.start()

            welcome = self.get_output()
            self.assertEqual(welcome, (
                "TODOs:\n"
                "\n"
                "\n"
                "> "
            ))

            self.send_input("add buy milk")
            self.send_input("quit")
            app_thread.join(timeout=1)

            while True:
                try:
                    self.get_output()
                except queue.Empty:
                    break
            app_thread = threading.Thread(
                target=TODOApp(
                    io=(self.fake_input, self.fake_output),
                    dbpath=tmpdirname
                ).run,
                daemon=True
            )
```

```
app_thread.start()

welcome = self.get_output()
self.assertEqual(welcome, (
    "TODOs:\n"
    "1. buy milk\n"
    "\n"
    "> "
))

self.send_input("quit")
app_thread.join(timeout=1)
```

First of all, our test makes a new temporary directory called tmpdirname, where we are going to save our database for the app under test. Then, as in the previous acceptance test, it starts the application in the background, pointing it to our fake I/O and the temporary path for the database. Once the app starts, we verify that, on first execution, it starts with an empty TODO list. Then we add one item to the app and we quit. At this point, we can restart the application again using the same exact database path, and check that the item we added is still there after the app restarts. Then we can just quit the app, as it did what we wanted to test.

Obviously, if we start our test suite, we already know that our new acceptance test is not going to pass. We haven't implemented the persistence of our todo items at all and our app doesn't even accept a dbpath argument:

```
$ python -m unittest discover -v
test_main (tests.test_acceptance.TestTODOAcceptance) ... ok
test_persistence (tests.test_acceptance.TestTODOAcceptance) ... ERROR

======================================================================
ERROR: test_persistence (tests.test_acceptance.TestTODOAcceptance)
----------------------------------------------------------------------
Traceback (most recent call last):
  File "/testingbook/03_tdd/02_codedesign/tests/test_acceptance.py", line
72, in test_persistence
    dbpath=tmpdirname
TypeError: __init__() got an unexpected keyword argument 'dbpath'

----------------------------------------------------------------------
Ran 2 tests in 0.004s

FAILED (errors=1)
```

Our next step is to move one layer below and start working on our implementation.

Thus the tests that we are going to write will get further away from the end user point of view that we used in the acceptance tests, and move toward describing what we want our inner implementation to be.

For this reason, we are going to create a separate directory for these tests so that they don't get confused with the higher-level tests that tell the story from the user's point of view. So inside our `tests` directory, we are going to create a subdirectory for unit tests.

Then, inside that directory, we are going to add a `test_todoapp.py` file to start reasoning about how we want to change our TODOApp object to support persistence:

```
└── tests
    ├── __init__.py
    ├── test_acceptance.py
    └── unit
        ├── __init__.py
        └── test_todoapp.py
```

Our `test_todoapp.py` file is going to start with a very simple test, one to verify that we can accept a database path for our TODO app and that if omitted, it should use the current directory:

```python
import unittest
import tempfile
from pathlib import Path

from todo.app import TODOApp


class TestTODOApp(unittest.TestCase):
    def test_default_dbpath(self):
        app = TODOApp()
        assert Path(".").resolve() == Path(app._dbpath).resolve()

    def test_accepts_dbpath(self):
        expected_path = Path(tempfile.gettempdir(), "something")
        app = TODOApp(dbpath=str(expected_path))
        assert expected_path == Path(app._dbpath)
```

Now we can forget for a little about our acceptance tests and focus on our unit tests. We are going to run them in isolation with the `-k unit` option to confirm that they fail as we expect, and we can move on to adding support for the `dbpath` to our object:

```
$ python -m unittest discover -k unit
EE
================================================================
```

```
ERROR: test_accepts_dbpath (tests.unit.test_todoapp.TestTODOApp)
----------------------------------------------------------------------
Traceback (most recent call last):
  File "/testingbook/03_tdd/02_codedesign/tests/unit/test_todoapp.py", line
12, in test_accepts_dbpath
    app = TODOApp(dbpath=str(expected_path))
TypeError: __init__() got an unexpected keyword argument 'dbpath'


======================================================================
ERROR: test_default_dbpath (tests.unit.test_todoapp.TestTODOApp)
----------------------------------------------------------------------
Traceback (most recent call last):
  File "/testingbook/03_tdd/02_codedesign/tests/unit/test_todoapp.py", line
9, in test_default_dbpath
    assert Path(".").resolve() == Path(app._dbpath).resolve()
AttributeError: 'TODOApp' object has no attribute '_dbpath'


----------------------------------------------------------------------
Ran 2 tests in 0.001s

FAILED (errors=2)
```

The −k option for unit tests only runs the tests that contain the provided substring, so it's going to identify only our tests inside the unit directory. It would obviously also run any tests that had unit in the name, but it's generally a convenient way to select some tests to run without having to remember in which exact directory they exist.

Now the implementation is fairly easy, we just want to make TODOApp able to remember where it has to save the database and have it always available as TODOApp._dbpath. So we are going to modify our TODOApp.__init__ to accept the extra argument and put it aside:

```
...

class TODOApp:
    def __init__(self,
                 io=(input, functools.partial(print, end="")),
                 dbpath=None):
        self._in, self._out = io
        self._quit = False
        self._entries = []
        self._dbpath = dbpath or "."

    ...
```

If we did this correctly, the tests for our implementation should now pass without issue:

```
$ python -m unittest discover -k unit -v
test_accepts_dbpath (tests.unit.test_todoapp.TestTODOApp) ... ok
test_default_dbpath (tests.unit.test_todoapp.TestTODOApp) ... ok


----------------------------------------------------------------------
Ran 2 tests in 0.002s

OK
```

And we can now look back to our acceptance test to find guidance about what to do next:

```
$ python -m unittest discover
.F..
======================================================================
FAIL: test_persistence (tests.test_acceptance.TestTODOAcceptance)
----------------------------------------------------------------------
Traceback (most recent call last):
  File "/tddbook/03_tdd/02_codedesign/tests/test_acceptance.py", line 108,
in test_persistence
    "TODOs:\n"
AssertionError: 'TODOs:\n\n\n> ' != 'TODOs:\n1. buy milk\n\n> '
  TODOs:
-
+ 1. buy milk
  >


----------------------------------------------------------------------
Ran 4 tests in 1.006s

FAILED (failures=1)
```

So, now our TODO application is able to start and accept the temporary database path. But it's not doing what we need. It's not saving anything into the database, so once restarted, the TODO list is still empty.

At this point, we need to go back to our unit tests and come up with a set of tests to drive the implementation of our persistence layer so that the data can be saved and loaded back.

Our first test should probably assess that TODOApp is able to load some save data. When we start thinking of our TestTODOApp.test_load test, it's easy to imagine the *Act* phase: it probably just wants to call a TODOApp.load method to load the data. The *Assert* phase too is also pretty obvious: TODOApp._entries should probably contain the same exact entries that we loaded.

But what about the *Arrange* phase? What are we going to store in the database so that we can load it back? Which database system are we going to use? And after a while we will probably move to the *"should we even care at all?"* question.

Does TODOApp have to care about how data is saved into the database?

Probably not... We should probably delegate that whole problem to another entity, and only make sure that TODOApp properly invokes that entity and does the right thing with the data provided by that entity:

```
...
from unittest.mock import Mock


class TestTODOApp(unittest.TestCase):
    ...

    def test_load(self):
        dbpath = Path(tempfile.gettempdir(), "something")
        dbmanager = Mock(
            load=Mock(return_value=["buy milk", "buy water"])
        )
        app = TODOApp(io=(Mock(return_value="quit"), Mock()),
                      dbpath=dbpath, dbmanager=dbmanager)
        app.run()

        dbmanager.load.assert_called_with(dbpath)
        assert app._entries == ["buy milk", "buy water"]
```

Our new TestTODOApp.test_load now tests this, provided dbmanager is in charge of loading/saving data. Our TODOApp is going to use it when it starts, and by virtue of calling dbmanager, it ends up with the todo entries that dbmanager loaded.

The test prepares a dbpath object for the sole purpose of checking that dbmanager is asked to load that specific path, then it makes a dbmanager that returns a canned response of two items when dbmanager.load(dbpath) is invoked. Once those two are in place, it prepares a TODOApp that has a dummy output and a stubbed input that make the app quit immediately.

Then, once the app is started through app.run(), we expect it to have called dbmanager and have loaded the two provided entries.

Now that we have a clearer understanding of what we want to do, we can go back to our TODOApp and write an implementation that satisfies our test. We are going to extend TODOApp to support dbmanager and we are going to modify TODOApp.run to load the existing data when the app is started:

```
class TODOApp:
    def __init__(self,
                    io=(input, functools.partial(print, end="")),
                    dbpath=None, dbmanager=None):
        self._in, self._out = io
        self._quit = False
        self._entries = []
        self._dbpath = dbpath or "."
        self._dbmanager = dbmanager

    def run(self):
        if self._dbmanager is not None:
            self._entries = self._dbmanager.load(self._dbpath)

        self._quit = False
        while not self._quit:
            self._out(self.prompt(self.items_list()))
            command = self._in()
            self._dispatch(command)

        self._out("bye!\n")
```

Is this enough to make our test pass? Let's find out:

```
$ python -m unittest discover -k unit -v
test_accepts_dbpath (tests.unit.test_todoapp.TestTODOApp) ... ok
test_default_dbpath (tests.unit.test_todoapp.TestTODOApp) ... ok
test_load (tests.unit.test_todoapp.TestTODOApp) ... ok

----------------------------------------------------------------------
Ran 3 tests in 0.002s

OK
```

It seems so, which means we achieved what we wanted. But there is something odd in our implementation. If TODOApp doesn't care about how data is loaded, why does it care where it is loaded from? The fact that you even need a path from which to load your data seems a concern of the loader. Maybe we can load data without a path? Maybe we can load things from remote resources that need a host and port instead of a path? That's something that only the loader can know.

So let's leverage our refactoring phase, as we made the tests pass, and change everything to just receive dbmanager. Whether that dbmanager needs a path, and whether that path is to a file, a directory, or a remote resource, is not something our app should care about.

First, we want to update the tests; instead of passing dbpath, we directly provide dbmanager itself. dbmanager will know the path. Let's also make a test for the case when no dbmanager is provided so that the app doesn't crash, but just disables persistency:

```
import unittest
from unittest.mock import Mock

from todo.app import TODOApp


class TestTODOApp(unittest.TestCase):
    def test_noloader(self):
        app = TODOApp(io=(Mock(return_value="quit"), Mock()),
                      dbmanager=None)

        app.run()

        assert app._entries == []

    def test_load(self):
        dbmanager = Mock(
            load=Mock(return_value=["buy milk", "buy water"])
        )
        app = TODOApp(io=(Mock(return_value="quit"), Mock()),
                      dbmanager=dbmanager)

        app.run()

        dbmanager.load.assert_called_with()
        assert app._entries == ["buy milk", "buy water"]
```

The first test_noloader test verifies that if there is no dbmanager, the app is still able to start, while test_load verifies that when dbmanager is used, the data that it provides is properly loaded by TODOApp.

We can now also throw away our test_accepts_dbpath and test_default_dbpath, as our TODOApp is no longer in charge of opening the database itself.

Do our newly refactored tests pass? Nope, not anymore:

```
$ python -m unittest discover -k unit -v
test_load (tests.unit.test_todoapp.TestTODOApp) ... FAIL
```

```
test_noloader (tests.unit.test_todoapp.TestTODOApp) ... ok


========================================================================
FAIL: test_load (tests.unit.test_todoapp.TestTODOApp)
------------------------------------------------------------------------
Traceback (most recent call last):
  File "/tddbook/03_tdd/02_codedesign/tests/unit/test_todoapp.py", line 29,
in test_load
    dbmanager.load.assert_called_with()
  File "/usr/lib/python3.7/unittest/mock.py", line 873, in
assert_called_with
    raise AssertionError(_error_message()) from cause
AssertionError: Expected call: load()
Actual call: load('.')


------------------------------------------------------------------------

Ran 2 tests in 0.002s

FAILED (failures=1)
```

Our mock expectation was violated. We expected `load` to be called with no argument, as `dbmanager` should already know where to load from, but instead, we received ".", which is the default `dbpath`.

Let's head back to our `TODOApp` and remove any reference to `dbpath`, thus removing the `dbpath` argument and the `self._dbpath` attribute:

```python
class TODOApp:
    def __init__(self,
                 io=(input, functools.partial(print, end="")),
                 dbmanager=None):
        self._in, self._out = io
        self._quit = False
        self._entries = []
        self._dbmanager = dbmanager

    def run(self):
        if self._dbmanager is not None:
            self._entries = self._dbmanager.load()

        self._quit = False
        while not self._quit:
            self._out(self.prompt(self.items_list()))
            command = self._in()
            self._dispatch(command)

        self._out("bye!\n")
```

Do our tests now pass? Yes! They do:

```
$ python -m unittest discover -k unit -v
test_load (tests.unit.test_todoapp.TestTODOApp) ... ok
test_noloader (tests.unit.test_todoapp.TestTODOApp) ... ok

----------------------------------------------------------------------
Ran 2 tests in 0.001s

OK
```

Now that we are happy with our implementation, we can go back to look for things to do.

When looking for things to do, guidance comes from our acceptance tests. If we run them right now they will probably crash because, in the end, we settled for an interface that is slightly different from the one we originally thought of:

```
$ python -m unittest discover
.E..
======================================================================
ERROR: test_persistence (tests.test_acceptance.TestTODOAcceptance)
----------------------------------------------------------------------
Traceback (most recent call last):
  File
"/home/amol/wrk/HandsOnTestDrivenDevelopmentPython/03_tdd/02_codedesign/tes
ts/test_acceptance.py", line 74, in test_persistence
    dbpath=tmpdirname,
TypeError: __init__() got an unexpected keyword argument 'dbpath'

----------------------------------------------------------------------
Ran 4 tests in 0.005s

FAILED (errors=1)
```

We don't receive `dbpath` anymore, but we want `dbmanager`. So let's update our test accordingly.

For now, we don't want to be too refined about our storage; we are just going to store things in a very simple storage system. Let's call this `BasicDB` and provide it to the app in our acceptance tests. They will load and save data from it:

```
...
import pathlib

...
from todo.db import BasicDB
```

```
class TestTODOAcceptance(unittest.TestCase):
    ...

    def test_persistence(self):
        with tempfile.TemporaryDirectory() as tmpdirname:
            app_thread = threading.Thread(
                target=TODOApp(
                    io=(self.fake_input, self.fake_output),
                    dbmanager=BasicDB(pathlib.Path(tmpdirname, "db"))
                ).run,
                daemon=True
            )
            app_thread.start()

        ...
```

Running our acceptance test now will tell us that the idea might look great, but we still have to implement `BasicDB`. So let's create a `tests/unit/test_basicdb.py` file and start reasoning how `BasicDB` should behave.

Our `TestBasicDB` tests are probably going to be for loading and saving data; for now, let's start with the loading one as that's what we are concerned about:

```
import pathlib
import unittest
from unittest import mock

from todo.db import BasicDB


class TestBasicDB(unittest.TestCase):
    def test_load(self):
        mock_file = mock.MagicMock(
            read=mock.Mock(return_value='["first", "second"]')
        )
        mock_file.__enter__.return_value = mock_file
        mock_opener = mock.Mock(return_value=mock_file)
        db = BasicDB(pathlib.Path("testdb"), _fileopener=mock_opener)
        loaded = db.load()

        self.assertEqual(loaded, ["first", "second"])
        self.assertEqual(
            mock_opener.call_args[0][0],
            pathlib.Path("testdb")
        )
        mock_file.read.assert_called_with()
```

We want our `BasicDB` to read/write data from a file, so we are going to use a `mock_file` object that fakes the Python behavior of a `file` object. When trying to read from it, it's going to return the content of our `BasicDB` with two sample entries.

`mock_file` is going to be what our `mock_opener` is going to return whenever `BasicDB` asks to open a new file. In practice, what we are trying to do is to make sure that `with mock_opener(ANY_PATH) as f:` will return our `mock_file`, so that from the point of view of `BasicDB`, there is no difference between using our `mock_opener` or the Python `open` function.

Once our stubbed file opener is available, we are going to create an instance of `BasicDB`, providing the stub opener as a replacement for the Python `open` function. The path we are going to provide to `BasicDB` for the storage of its database doesn't really matter at this point as it will always return `mock_file`, but we will still be checking that the opener was called with the expected path.

The real core of our test is the call to `db.load()`, where we are going to ask `BasicDB` to load the data from `mock_file`. Then we can confirm that the data we expected was loaded and that it was loaded the way we would expect, by actually opening the file and reading its content.

In practice, we decided that `BasicDB(path).load()` will be the way we plan to load the data in `BasicDB`.

Now that we've set our expectations clearly and have a better idea of what we want to build, we can try to work on an implementation that could satisfy the interface we imagined.

The first step is creating our `src/todo/db.py` module, as that's where we imagined we would be importing `BasicDB` from while writing our test (see the `from todo.db import BasicDB` line at the top of our test file).

Then we are going to make a `BasicDB` class that accepts the file path to save/load data to/from, and an optional opener so that we can replace the default one with other alternative implementations. For the goal of making clear that the opener is mostly meant for testing, we are going to flag it as an internal detail, prefixing its name with an underscore:

```
class BasicDB:
    def __init__(self, path, _fileopener=open):
        self._path = path
        self._fileopener = _fileopener
```

Will this make our tests pass? I doubt it will – it still does nothing, so let's cycle back to our tests to see which parts of the `BasicDB` interface we have to implement:

```
$ python -m unittest discover -k unit -v
test_load (tests.unit.test_basicdb.TestBasicDB) ... ERROR
test_load (tests.unit.test_todoapp.TestTODOApp) ... ok
test_noloader (tests.unit.test_todoapp.TestTODOApp) ... ok


========================================================================
ERROR: test_load (tests.unit.test_basicdb.TestBasicDB)
------------------------------------------------------------------------
Traceback (most recent call last):
  File "/tddbook/03_tdd/02_codedesign/tests/unit/test_basicdb.py", line 18,
in test_load
    loaded = db.load()
AttributeError: 'BasicDB' object has no attribute 'load'


------------------------------------------------------------------------
Ran 3 tests in 0.002s

FAILED (errors=1)
```

OK, it seems we now want to move to the implementation of `BasicDB.load`.

The implementation feels pretty straightforward: we open a file that should contain a list of strings. Let's just read the file content and parse the list definition:

```
class BasicDB:
    def __init__(self, path, _fileopener=open):
        self._path = path
        self._fileopener = _fileopener

    def load(self):
        with self._fileopener(self._path, "r", encoding="utf-8") as f:
            txt = f.read()
        return eval(txt)
```

Does this make our tests happy? Are we really able to load the items stored in `BasicDB`? Let's find out:

```
$ python -m unittest discover -k unit -v
test_load (tests.unit.test_basicdb.TestBasicDB) ... ok
test_load (tests.unit.test_todoapp.TestTODOApp) ... ok
test_noloader (tests.unit.test_todoapp.TestTODOApp) ... ok


----------------------------------------------------------------------
Ran 3 tests in 0.002s

OK
```

It seems so – our `BasicDB` test was able to load the content and fetch back the two items.

 For anyone wondering about the usage of `eval`, please bear with the example for a little while. We are going to replace it pretty soon and make clear that using it is never a good idea. But it was a convenient way to simulate the bug we are going to fix in the dedicated *Preventing regressions* section.

All our unit tests now pass, so we are a bit at a loss about where we were and what we wanted to do next. Whenever we are unsure about our next step, the acceptance tests should guide us on how far we are from the feature we want to provide for our users. So let's go back to our acceptance test and see what we still have to do:

```
$ python -m unittest discover -k acceptance
...
FileNotFoundError: [Errno 2] No such file or directory:
'/tmp/tmpcug9zvsw/db'
```

Uh, we forgot that when we start the application the first time, our `BasicDB` is empty; actually, it doesn't exist at all. So there is nothing we can load. Thus we have to go back to our unit tests and write one to ensure that when the opened file doesn't exist, we do actually return an empty list of todo items.

Back to our `tests/unit/test_basicdb.py` file, we are going to add a new `test_missing_load` test:

```
...
class TestBasicDB(unittest.TestCase):
    ...

    def test_missing_load(self):
        mock_opener = mock.Mock(side_effect=FileNotFoundError)
```

```
        db = BasicDB(pathlib.Path("testdb"), _fileopener=mock_opener)
        loaded = db.load()

        self.assertEqual(loaded, [])
        self.assertEqual(
            mock_opener.call_args[0][0],
            pathlib.Path("testdb")
        )
```

This new test is just going to throw `FileNotFoundError` every time `BasicDB` tries to read the data. This simulates the case where we would try to open a nonexistent database.

As expected, our test is going to fail with `FileNotFoundError` as we haven't handled it yet:

```
$ python -m unittest discover -k unit -v
test_load (tests.unit.test_basicdb.TestBasicDB) ... ok
test_missing_load (tests.unit.test_basicdb.TestBasicDB) ... ERROR
test_load (tests.unit.test_todoapp.TestTODOApp) ... ok
test_noloader (tests.unit.test_todoapp.TestTODOApp) ... ok


======================================================================
ERROR: test_missing_load (tests.unit.test_basicdb.TestBasicDB)
----------------------------------------------------------------------
Traceback (most recent call last):
  File "/tddbook/03_tdd/02_codedesign/tests/unit/test_basicdb.py", line 31,
in test_missing_load
    loaded = db.load()
  File "/tddbook/03_tdd/02_codedesign/src/todo/db.py", line 9, in load
    with self._fileopener(self._path, "r", encoding="utf-8") as f:
  File "/usr/lib/python3.7/unittest/mock.py", line 1011, in __call__
    return _mock_self._mock_call(*args, **kwargs)
  File "/usr/lib/python3.7/unittest/mock.py", line 1071, in _mock_call
    raise effect
FileNotFoundError


----------------------------------------------------------------------
Ran 4 tests in 0.003s

FAILED (errors=1)
```

But we can easily modify our `BasicDB.load` method to handle such a case and return an empty list of todo items:

```
class BasicDB:
    def __init__(self, path, _fileopener=open):
        self._path = path
        self._fileopener = _fileopener
```

```
    def load(self):
        try:
            with self._fileopener(self._path, "r",
                                  encoding="utf-8") as f:
                txt = f.read()
            return eval(txt)
        except FileNotFoundError:
            return []
```

At this point, if we got it right, our unit tests should all pass:

```
$ python -m unittest discover -k unit -v
test_load (tests.unit.test_basicdb.TestBasicDB) ... ok
test_missing_load (tests.unit.test_basicdb.TestBasicDB) ... ok
test_load (tests.unit.test_todoapp.TestTODOApp) ... ok
test_noloader (tests.unit.test_todoapp.TestTODOApp) ... ok


----------------------------------------------------------------------
Ran 4 tests in 0.002s

OK
```

Given that we were looking for our next step a few minutes ago, we should probably head back to our acceptance tests and check where we were. Running our acceptance tests again will show that this time, we were able to start the application correctly (that is, it doesn't crash anymore on missing files), but that on adding a new item and restarting the app, it didn't persist the addition:

```
$ python -m unittest discover -k acceptance
.F
======================================================================
FAIL: test_persistence (tests.test_acceptance.TestTODOAcceptance)
----------------------------------------------------------------------
Traceback (most recent call last):
  File "/tddbook/03_tdd/02_codedesign/tests/test_acceptance.py", line 110,
in test_persistence
    "TODOs:\n"
AssertionError: 'TODOs:\n\n\n> ' != 'TODOs:\n1. buy milk\n\n> '
  TODOs:
-
+ 1. buy milk
  >


----------------------------------------------------------------------
Ran 2 tests in 1.006s

FAILED (failures=1)
```

The `buy milk` item is not where we expected it to be after reloading the application, which makes sense, as we never actually implemented any support for saving the current todo items when we exit the application. So while we are probably able to load back a list of items, we never save one.

This means we want to extend our `TODOApp` to save the current list of todo items before exiting.

So let's add a `test_save` test to our `tests/unit/tests_todoapp.py` tests to make clear what we want to achieve.

We just want the application to start with some entries and make sure that when it quits, the app asks `dbmanager` to save them. This means that if there was any change made to our list of TODOs, it gets recorded:

```python
class TestTODOApp(unittest.TestCase):
    ...

    def test_save(self):
        dbmanager = Mock(
            load=Mock(return_value=["buy milk", "buy water"]),
            save=Mock()
        )

        app = TODOApp(io=(Mock(return_value="quit"), Mock()),
                      dbmanager=dbmanager)
        app.run()

        dbmanager.save.assert_called_with(["buy milk", "buy water"])
```

This test will obviously fail because we haven't yet used the `dbmanager` from `TODOApp` to save anything:

```
$ python -m unittest discover -k unit -v
test_load (tests.unit.test_basicdb.TestBasicDB) ... ok
test_missing_load (tests.unit.test_basicdb.TestBasicDB) ... ok
test_load (tests.unit.test_todoapp.TestTODOApp) ... ok
test_noloader (tests.unit.test_todoapp.TestTODOApp) ... ok
test_save (tests.unit.test_todoapp.TestTODOApp) ... FAIL

======================================================================
FAIL: test_save (tests.unit.test_todoapp.TestTODOApp)
----------------------------------------------------------------------
Traceback (most recent call last):
  File "/tddbook/03_tdd/02_codedesign/tests/unit/test_todoapp.py", line 39,
in test_save
    dbmanager.save.assert_called_with(["buy milk", "buy water"])
```

```
   File "/usr/lib/python3.7/unittest/mock.py", line 864, in
assert_called_with
     raise AssertionError('Expected call: %s\nNot called' % (expected,))
AssertionError: Expected call: save(['buy milk', 'buy water'])
Not called

----------------------------------------------------------------------
Ran 5 tests in 0.003s

FAILED (failures=1)
```

So, let's go to our `TODOApp.run` method and extend it to call `dbmanager.save()` before exiting:

```python
class TODOApp:
    ...

    def run(self):
        if self._dbmanager is not None:
            self._entries = self._dbmanager.load()

        self._quit = False
        while not self._quit:
            self._out(self.prompt(self.items_list()))
            command = self._in()
            self._dispatch(command)

        if self._dbmanager is not None:
            self._dbmanager.save(self._entries)
        self._out("bye!\n")
```

That's all we need to make our test pass. Our `TODOApp` now takes care of saving the entries and it's up to the provided `dbmanager` to do the right thing with them:

```
$ python -m unittest discover -k unit -v
test_load (tests.unit.test_basicdb.TestBasicDB) ... ok
test_missing_load (tests.unit.test_basicdb.TestBasicDB) ... ok
test_load (tests.unit.test_todoapp.TestTODOApp) ... ok
test_noloader (tests.unit.test_todoapp.TestTODOApp) ... ok
test_save (tests.unit.test_todoapp.TestTODOApp) ... ok

----------------------------------------------------------------------
Ran 5 tests in 0.002s

OK
```

Are we done? Not yet – TODOApp is now doing its job, but a quick run of our acceptance test will point out that dbmanager doesn't know what we are talking about:

```
$ python -m unittest discover -k acceptance
.Exception in thread Thread-2:
Traceback (most recent call last):
  File "/usr/lib/python3.7/threading.py", line 926, in _bootstrap_inner
    self.run()
  File "/usr/lib/python3.7/threading.py", line 870, in run
    self._target(*self._args, **self._kwargs)
  File "/tddbook/03_tdd/02_codedesign/src/todo/app.py", line 24, in run
    self._dbmanager.save(self._entries)
AttributeError: 'BasicDB' object has no attribute 'save'
```

Back to our tests/unit/test_basicdb.py file, we are going to add a test_save test to confirm that BasicDB does actually want to save the list of provided items:

```
class TestBasicDB(unittest.TestCase):
    ...

    def test_save(self):
        mock_file = mock.MagicMock(write=mock.Mock())
        mock_file.__enter__.return_value = mock_file
        mock_opener = mock.Mock(return_value=mock_file)
        db = BasicDB(pathlib.Path("testdb"), _fileopener=mock_opener)
        loaded = db.save(["first", "second"])

        self.assertEqual(
            mock_opener.call_args[0][0:2],
            (pathlib.Path("testdb"), "w+")
        )
        mock_file.write.assert_called_with('["first", "second"]')
```

The test just verifies that when BasicDB.save is called, it opens the target file in write mode and it tries to write into it the list of values.

To satisfy our test, we are going to implement a BasicDB.save method that converts the list of entries to its string representation, replaces single quotes with double quotes so that we save them in a format that is compatible with JSON, and saves it back:

```
class BasicDB:
    ...

    def save(self, values):
        with self._fileopener(self._path, "w+", encoding="utf-8") as f:
            f.write(repr(values).replace("'", '"'))
```

If we did everything correctly, our unit tests should now be able to pass:

```
$ python -m unittest discover -k unit -v
test_load (tests.unit.test_basicdb.TestBasicDB) ... ok
test_missing_load (tests.unit.test_basicdb.TestBasicDB) ... ok
test_save (tests.unit.test_basicdb.TestBasicDB) ... ok
test_load (tests.unit.test_todoapp.TestTODOApp) ... ok
test_noloader (tests.unit.test_todoapp.TestTODOApp) ... ok
test_save (tests.unit.test_todoapp.TestTODOApp) ... ok


----------------------------------------------------------------------
Ran 6 tests in 0.003s

OK
```

We implemented everything that we wanted and we provided the last piece that our acceptance test was complaining about, which can be easily confirmed by going back to our acceptance tests and verifying that the software is now completed:

```
$ python -m unittest discover -k acceptance
..
----------------------------------------------------------------------
Ran 2 tests in 1.006s

OK
```

Great! Our app is now fully functional.

We just want to tweak our `src/todo/_main__.py` file so that when we start the app from the command line, we start it with `dbmanager` and thus with persistence enabled by default:

```python
from .app import TODOApp
from .db import BasicDB

TODOApp(dbmanager=BasicDB("todo.data")).run()
```

Starting the application, adding an entry, and then restarting it will now properly preserve the entry across the two runs:

```
$ python -m todo
TODOs:


> add buy milk
TODOs:
1. buy milk
```

```
> quit
bye!

$ python -m todo
TODOs:
1. buy milk

> quit
bye!
```

Before ending our day with a sense of satisfaction from our newly built application, we want to make sure we remember to install the new release of our favorite Linux distribution.

As we just made a great TODO application, let's add an entry to it:

```
$ python -m todo
TODOs:
1. buy milk

> add install "Focal Fossa"
TODOs:
1. buy milk
2. install "Focal Fossa"

> quit
bye!
```

Sadly, the morning after, we open our TODO application to look at what we have to do, and surprise, surprise, we are welcomed by a major crash in our application:

```
$ python -m todo
Traceback (most recent call last):
  File "/usr/lib/python3.7/runpy.py", line 193, in _run_module_as_main
    "__main__", mod_spec)
  File "/usr/lib/python3.7/runpy.py", line 85, in _run_code
    exec(code, run_globals)
  File "/tddbook/03_tdd/02_codedesign/src/todo/__main__.py", line 4, in
<module>
    TODOApp(dbmanager=BasicDB("todo.data")).run()
  File "/tddbook/03_tdd/02_codedesign/src/todo/app.py", line 15, in run
    self._entries = self._dbmanager.load()
  File "/tddbook/03_tdd/02_codedesign/src/todo/db.py", line 12, in load
    return eval(txt)
  File "<string>", line 1
    ["buy milk", "install "Focal Fossa""]
                              ^
SyntaxError: invalid syntax
```

Our data is unable to load due to an issue in the `BasicDB` persistence layer, and we will have to fix our bug if we ever want to be able to use our TODO application. This is actually great because TDD has a best practice that allows us to tackle these bugs. Let's introduce regression tests.

Preventing regressions

Tests are not only used to drive our application design and our code design, but also drive our research and the debugging of the issues that our application faces.

Whenever we face any kind of error, bug, or crash, our fixing process should start with writing a **regression test** – a test whose purpose is to reproduce the same exact issue we are facing.

Having a regression test in place will prevent that bug from happening again in the future, even if someone refactors some of the code or replaces the implementation. That's not all a test can do – once we've written a test that reproduces our issue, we will be able to more easily debug the issue and see what's going on in a fully controlled and isolated environment such as a test suite.

As our application crashed trying to load our database, we are going to write a test for it and see what the problem is.

The first step is writing a test that reproduces the same exact steps that the user did to trigger the condition, so we are going to write a test in `tests/test_regressions.py` that is going to reproduce our most recent user sessions in the application.

Our first goal is to be able to reproduce the issue. To do so, we are going to use the setup that is most similar to that in the real world. So we are going to reuse the setup code from our acceptance tests and create a `TestRegressions` class:

```
import unittest
import threading
import queue
import tempfile
import pathlib

from todo.app import TODOApp
from todo.db import BasicDB


class TestRegressions(unittest.TestCase):
    def setUp(self):
```

```
self.inputs = queue.Queue()
self.outputs = queue.Queue()

self.fake_output = lambda txt: self.outputs.put(txt)
self.fake_input = lambda: self.inputs.get()

self.get_output = lambda: self.outputs.get(timeout=1)
self.send_input = lambda cmd: self.inputs.put(cmd)
```

This is the same exact setUp code we had in our acceptance tests for fake I/O. We could inherit from the same base class or use a mixin to provide the setup of our fake I/O, but here we just copied those same few lines of code.

Then we are going to add a test_os_release method that reproduces exactly what happened in our real usage session:

```
def test_os_release(self):
    with tempfile.TemporaryDirectory() as tmpdirname:
        app_thread = threading.Thread(
            target=TODOApp(
                io=(self.fake_input, self.fake_output),
                dbmanager=BasicDB(pathlib.Path(tmpdirname, "db"))
            ).run,
            daemon=True
        )
        app_thread.start()
        self.get_output()

        self.send_input("add buy milk")
        self.send_input('add "Focal Fossa"')
        self.send_input("quit")
        app_thread.join(timeout=1)

        while True:
            try:
                self.get_output()
            except queue.Empty:
                break
        app_thread = threading.Thread(
            target=TODOApp(
                io=(self.fake_input, self.fake_output),
                dbmanager=BasicDB(pathlib.Path(tmpdirname, "db"))
            ).run,
            daemon=True
        )
        app_thread.start()
        self.get_output()
```

First, we start the application, then we add a note to buy milk, install the `Focal Fossa` release, and then we quit. Subsequently, we just restart the application.

If we run our test, it should reproduce the same exact steps that happened in our software and thus trigger the same exact crash:

```
$ python -m unittest discover -k regression
Exception in thread Thread-2:
Traceback (most recent call last):
  File "/usr/lib/python3.8/threading.py", line 932, in _bootstrap_inner
    self.run()
  File "/usr/lib/python3.8/threading.py", line 870, in run
    self._target(*self._args, **self._kwargs)
  File "/tddbook/03_tdd/03_regression/src/todo/app.py", line 15, in run
    self._entries = self._dbmanager.load()
  File "/tddbook/03_tdd/03_regression/src/todo/db.py", line 12, in load
    return eval(txt)
  File "<string>", line 1
    ["buy milk", "install "Focal Fossa""]
                          ^
SyntaxError: invalid syntax
```

OK, the crash is there and it's the same exact traceback. So we were able to reproduce the issue! Our next step is to isolate the issue to find what really causes it and which part of our system is involved in the problem itself.

To do so, we are going to move from a test that really runs the application to a simpler one that does not involve the whole machinery and I/O support. Let's see whether we can reproduce the issue by replacing our fairly long and complete `TestRegressions` class with one that just starts the application with a stubbed set of inputs and then restarts it:

```
import unittest
from unittest import mock
import tempfile
import pathlib

from todo.app import TODOApp
from todo.db import BasicDB


class TestRegressions(unittest.TestCase):
    def test_os_release(self):
        with tempfile.TemporaryDirectory() as tmpdirname:
            app = TODOApp(
                io=(mock.Mock(side_effect=[
                    "add buy milk",
                    'add install "Focal Fossa"',
```

```
                "quit"
            ]), mock.Mock()),
            dbmanager=BasicDB(pathlib.Path(tmpdirname, "db"))
        )
        app.run()

        restarted_app = TODOApp(
            io=(mock.Mock(side_effect="quit"), mock.Mock()),
            dbmanager=BasicDB(pathlib.Path(tmpdirname, "db"))
        )
        restarted_app.run()
```

If we rerun our regression tests, we are luckily going to see that it still fails as before:

```
$ python -m unittest discover -k regression
E
========================================================================
ERROR: test_os_release (tests.test_regressions.TestRegressions)
------------------------------------------------------------------------
Traceback (most recent call last):
  File "/tddbook/03_tdd/03_regression/tests/test_regressions.py", line 27,
in test_os_release
    restarted_app.run()
  File "/tddbook/03_tdd/03_regression/src/todo/app.py", line 15, in run
    self._entries = self._dbmanager.load()
  File "/tddbook/03_tdd/03_regression/src/todo/db.py", line 12, in load
    return eval(txt)
  File "<string>", line 1
    ["buy milk", "install "Focal Fossa""]
                          ^
SyntaxError: invalid syntax


------------------------------------------------------------------
Ran 1 test in 0.003s

FAILED (errors=1)
```

This helped us confirm that the I/O doesn't really matter and that running the application for real is not involved in causing our bug. We greatly reduced the scope of the involved entities to just TODOApp and BasicDB objects.

There is still the filesystem involved; does that matter? Is it a problem with the fact that we are reading and writing files?

To check that, let's move forward further and get rid of the filesystem too. We can use an opener that provides an in-memory file instead of a real one so that where we write doesn't matter anymore:

```python
import unittest
from unittest import mock
import io

from todo.app import TODOApp
from todo.db import BasicDB


class TestRegressions(unittest.TestCase):
    def test_os_release(self):
        fakefile = io.StringIO()
        fakefile.close = mock.Mock()

        app = TODOApp(
            io=(mock.Mock(side_effect=[
                "add buy milk",
                'add install "Focal Fossa"',
                "quit"
            ]), mock.Mock()),
            dbmanager=BasicDB(None, _fileopener=mock.Mock(
                side_effect=[FileNotFoundError, fakefile]
            ))
        )
        app.run()

        # rollback the file. So that the application, restarting,
        # can read the new data that we wrote.
        fakefile.seek(0)

        restarted_app = TODOApp(
            io=(mock.Mock(return_value="quit"), mock.Mock()),
            dbmanager=BasicDB(None, _fileopener=mock.Mock(
                return_value=fakefile
            ))
        )
        restarted_app.run()
```

Our test now creates an `io.StringIO` instance instead of using a real file, so it doesn't depend anymore on a real disk. We replaced the standard `io.StringIO.close()` method with a dummy one, so that the file never gets closed and we can read it again. Otherwise, after it's used for the first time it will be lost forever.

Then we started the application with a `_fileopener` that firstly triggers `FileNotFoundError`, causing the application to start with an empty todo list, and secondly returns the fake file so that the data gets saved to the fake file. The same fake file, from which the application once restarted, will read the todo items.

Rerunning our regression test will confirm that we are still able to reproduce the same exact issue, and thus our test is still valid:

```
$ python -m unittest discover -k regression
E
========================================================================
ERROR: test_os_release (tests.test_regressions.TestRegressions)
------------------------------------------------------------------------
Traceback (most recent call last):
  File "/tddbook/03_tdd/03_regression/tests/test_regressions.py", line 36,
in test_os_release
    restarted_app.run()
  File "/tddbook/03_tdd/03_regression/src/todo/app.py", line 15, in run
    self._entries = self._dbmanager.load()
  File "/tddbook/03_tdd/03_regression/src/todo/db.py", line 12, in load
    return eval(txt)
  File "<string>", line 1
    ["buy milk", "install "Focal Fossa""]
                          ^
SyntaxError: invalid syntax


------------------------------------------------------------------------
Ran 1 test in 0.002s

FAILED (errors=1)
```

OK, we removed every interaction with the outer world. We know that our problem can be reproduced solely with `TODOApp` and `BasicDB`. What else can we try to remove from the equation to further reduce the area where our issue might live and identify the minimum system components necessary to reproduce our issue?

Our issue crashes in `BasicDB.load()`, so there is a high chance that it's caused by loading back the data that we saved. So let's get rid of `TODOApp` and try to directly save and load back our list of two items.

Our final version of the test is fairly minimal and has isolated `BasicDB` on its own:

```
class TestRegressions(unittest.TestCase):
    def test_os_release(self):
        fakefile = io.StringIO()
        fakefile.close = mock.Mock()
```

```
            data = ["buy milk", 'install "Focal Fossa"']

            dbmanager = BasicDB(None, _fileopener=mock.Mock(
                return_value=fakefile
            ))

            dbmanager.save(data)
            fakefile.seek(0)
            loaded_data = dbmanager.load()
            self.assertEqual(loaded_data, data)
```

Running our test does indeed fail with the same exact error that we had before:

```
$ python -m unittest discover -k regression
E
========================================================================
ERROR: test_os_release (tests.test_regressions.TestRegressions)
------------------------------------------------------------------------
Traceback (most recent call last):
  File
"/home/amol/wrk/HandsOnTestDrivenDevelopmentPython/03_tdd/03_regression/tes
ts/test_regressions.py", line 22, in test_os_release
    loaded_data = dbmanager.load()
  File
"/home/amol/wrk/HandsOnTestDrivenDevelopmentPython/03_tdd/03_regression/src
/todo/db.py", line 12, in load
    return eval(txt)
  File "<string>", line 1
    ["buy milk", "install "Focal Fossa""]
                          ^
SyntaxError: invalid syntax


------------------------------------------------------------------------
Ran 1 test in 0.001s

FAILED (errors=1)
```

So we were able to get a test involving the minimum possible number of entities in isolation to reproduce our issue. Only BasicDB is in use in our test, so we now know for sure that that's where our issue lies.

Our issue is due to the fact that we tried to save and load data in JSON format, relying on the fact that the Python syntax for arrays of strings is nearly the same as JSON. Thus using repr and eval could work to generate the JSON and load it back.

Sadly, that was a pretty terrible idea that we put in place for the sole purpose of reproducing this issue. Evaluating user inputs is generally a big security hole.

If instead of `install "Focal Fossa"`, we wrote `"] + [print("hello")] + ["` as our todo item, that would have resulted in our `TODOApp` executing the Python `print` function when loading back todo items (because what we saved was `["buy milk", ""] + [print("hello")] + [""]`) and instead of `print`, we could have forced the app to do anything when loading back the todo items.

`eval` should never be used with input that comes from users, so let's replace our `BasicDB` implementation with one that uses the `json` module:

```
import json


class BasicDB:
    def __init__(self, path, _fileopener=open):
        self._path = path
        self._fileopener = _fileopener

    def load(self):
        try:
            with self._fileopener(self._path, "r",
                                  encoding="utf-8") as f:
                return json.load(f)
        except FileNotFoundError:
            return []

    def save(self, values):
        with self._fileopener(self._path, "w+", encoding="utf-8") as f:
            f.write(json.dumps(values))
```

The only part we changed in `BasicDB.load` is that instead of using `eval`, we now use `json.load`, and in `BasicDB.save`, instead of `repr` we use `json.dumps`.

This uses the JSON module to save and load our data, removing the risk of malicious code execution.

If we did everything correctly, our test for the bug should finally pass, while our application continues to pass all other existing tests as well:

```
$ python -m unittest discover -v
test_main (tests.test_acceptance.TestTODOAcceptance) ... ok
test_persistence (tests.test_acceptance.TestTODOAcceptance) ... ok
test_os_release (tests.test_regressions.TestRegressions) ... ok
test_load (tests.unit.test_basicdb.TestBasicDB) ... ok
test_missing_load (tests.unit.test_basicdb.TestBasicDB) ... ok
test_save (tests.unit.test_basicdb.TestBasicDB) ... ok
test_load (tests.unit.test_todoapp.TestTODOApp) ... ok
```

```
test_noloader (tests.unit.test_todoapp.TestTODOApp) ... ok
test_save (tests.unit.test_todoapp.TestTODOApp) ... ok


----------------------------------------------------------------------
Ran 9 tests in 1.015s

OK
```

It seems we succeeded! We identified the bug, fixed it, and now have a test preventing the same bug from happening again.

I hope the benefit of starting any bug-and-issue resolution by first writing a test that reproduces the issue itself is clear. Not only does it prevent the issue from happening again in the future, but it also allows you to isolate the system where the bug is happening, design a fix, and make sure you actually fix the right bug.

Summary

We saw how acceptance tests can be used to make clear what we want to build and guide us step by step through what we have to build next, while lower-level tests, such as unit and integration tests, can be used to tell us how we want to build it and how we want the various pieces to work together.

In this case, our application was fairly small, so we used the acceptance test to verify the integration of our pieces. However, in the real world, as we grow the various parts of our infrastructure, we will have to introduce tests to confirm they are able to work together and the reason is their intercommunication protocol.

Once we found a bug, we also saw how regression tests can help us design fixes and how they can prevent the same bug from happening again in the long term.

During any stage of software development, the *Design, Implementation, and Maintenance* workflow helps us better understand what we are trying to do and thus get the right software, code, and bug fixes in place.

So far, we've worked with fairly small test suites, but the average real-world software has thousands of tests, so particular attention to how we organize will be essential to a test suite we feel we can rely on. In the next chapter, we are thus going to see how to scale test suites when the number of tests becomes hard to manage and the time it takes to run the test suite gets too long to run it all continuously.

4
Scaling the Test Suite

Writing one test is easy; writing thousands of tests, maintaining them, and ensuring they don't become a burden for development and the team is hard. Let's dive into some tools and best practices that help us define our test suite and keep it in shape.

To support the concepts in this chapter, we are going to use the test suite written for our Chat application in Chapter 2, *Test Doubles with a Chat Application*. We are going to see how to scale it as the application gets bigger and the tests get slower, and how to organize it in a way that can serve us in the long term.

In this chapter, we will cover the following topics:

- Scaling tests
- Working with multiple suites
- Carrying out performance testing
- Enabling continuous integration

Technical requirements

A working Python interpreter and a GitHub.com account are required to work through the examples in this chapter.

The examples we'll work through have been written using Python 3.7, but should work with most modern Python versions.

The source code for the examples in this chapter can be found on GitHub at https://github.com/PacktPublishing/Crafting-Test-Driven-Software-with-Python /tree/main/Chapter04

Scaling tests

When we started our Chat application in Chapter 2, *Test Doubles with a Chat Application*, the whole code base was contained in a single Python module. This module mixed both the application itself, the test suite, and the fakes that we needed for the test suite.

While that process fits well for the experimentation and hacking phase, it's not convenient for the long term. As we already saw in Chapter 3, *Test-Driven Development while Creating a TODO List*, it's possible to split tests into multiple files and directories and keep them separated from our application code.

As our project grows, the first step is to split our test suite from our code base. We are going to use the src directory for the code base and the tests directory for the test suite. The src directory in this case will contain the chat package, which contains the modules for the client and server code:

```
.
├── src
│   ├── chat
│   │   ├── client.py
│   │   ├── __init__.py
│   │   └── server.py
│   └── setup.py
```

The src/chat/client.py file will contain the previous Connection and ChatClient classes, while in src/chat/server.py we are going to put the new_chat_server function.

We also provide a very minimal src/setup.py to allow installation of the chat package:

```
from setuptools import setup

setup(name='chat', packages=['chat'])
```

Now that we can install the chat package through pip install -e ./src and then use any class within it through import chat, our tests can be moved anywhere; they no longer need to be in the same directory of the files they need to test.

Thus we can create a `tests` directory and gather all our tests there. As we had three different test classes (`TestChatAcceptance`, `TestChatClient`, and `TestConnection`), we are going to split our tests into three dedicated files. This way, while we work, we can run only tests relevant to the part we are modifying:

```
└── tests
    ├── __init__.py
    ├── test_chat.py
    ├── test_client.py
    └── test_connection.py
```

The only required changes to the tests we made in Chapter 2, *Test Doubles with a Chat Application,* are to make sure that we add proper imports to get our classes (for example, `from chat.client import ChatClient`). Once those are in place, our test suite should be able to run exactly as it used to:

```
$ python -m unittest discover -v
test_message_exchange (tests.test_chat.TestChatAcceptance) ... ok
test_client_connection (tests.test_client.TestChatClient) ... ok
test_client_fetch_messages (tests.test_client.TestChatClient) ... ok
test_nickname (tests.test_client.TestChatClient) ... ok
test_send_message (tests.test_client.TestChatClient) ... ok
test_broadcast (tests.test_connection.TestConnection) ... ok

----------------------------------------------------------------------
Ran 6 tests in 0.607s

OK
```

In a **Test-Driven Development** (**TDD**) approach, the test suite is something we will be able to run frequently and quickly to verify the work we are doing, but in a real-world application, test suites tend to become big and slow and can take minutes or hours to run.

For example, we might decide to grow our test suite further. Right now, we only have a test to verify that two users can exchange a message, but we have not verified that when multiple users are involved, we still see messages from all of them, and that each connected user sees the same exact messages.

To do so, we can add a new `TestChatMultiUser` test case to our `tests/test_chat.py` tests to verify that we can see the messages sent by all users connected to the chat:

```
class TestChatMultiUser(unittest.TestCase):
    def test_many_users(self):
        with new_chat_server() as srv:
            firstUser = ChatClient("John Doe")
```

```
for uid in range(5):
    moreuser = ChatClient(f"User {uid}")
    moreuser.send_message("Hello!")

messages = firstUser.fetch_messages()
assert len(messages) == 5
```

The `test_many_users` test connects to the chat as `firstUser` and then adds five more users to the chat and sends a new message from each of them. At the end of the test, `firstUser` should be able to see all five messages sent by the other users.

To go further, we could also add a `test_multiple_readers` test that verifies that all users in the chat see the same exact messages:

```
def test_multiple_readers(self):
    with new_chat_server() as srv:
        user1 = ChatClient("John Doe")
        user2 = ChatClient("User 2")
        user3 = ChatClient("User 3")

        user1.send_message("Hi all")
        user2.send_message("Hello World")
        user3.send_message("Hi")

        user1_messages = user1.fetch_messages()
        user2_messages = user2.fetch_messages()
        self.assertEqual(user1_messages, user2_messages)
```

In this case, we have three users joining the chat, each of them sending a message, and then we verified that both `user1` and `user2` see the same exact messages.

Through these two tests, we confirmed that our chat works as expected even when multiple users are inside the chat. If we receive messages from different users, we will see all messages, and all users will see the same exact messages. The side effect of the additional confidence that we now have in our chat is that our test suite has become far slower:

```
$ python -m unittest discover -v -k e2e -k unit
test_message_exchange (tests.test_chat.TestChatAcceptance) ... ok
test_many_users (tests.test_chat.TestChatMultiUser) ... ok
test_multiple_readers (tests.test_chat.TestChatMultiUser) ... ok
test_client_connection (tests.test_client.TestChatClient) ... ok
test_client_fetch_messages (tests.test_client.TestChatClient) ... ok
test_nickname (tests.test_client.TestChatClient) ... ok
test_send_message (tests.test_client.TestChatClient) ... ok
test_broadcast (tests.test_connection.TestConnection) ... ok

-------------------------------------------------------------------
```

```
Ran 8 tests in 3.589s
```

```
OK
```

From less than a second that it took previously to run our tests, we went to nearly 4 seconds.

As we grow our chat further, we are surely going to add more features, and more features will require more tests. Our test suite will become too slow and inconvenient to run as it will quickly reach minutes of time per run. Anything that runs in more than a few seconds is something that we are going to start running infrequently, thus moving further away from the benefits of a test-driven approach.

But we might argue that there will be kinds of tests that are always going to take a long time to run because they are slow by nature due to what they do (for example, performance tests), so what can we do to improve the situation?

A good first step is to make sure tests are properly spread out in groups that make their purpose and expected runtime clear.

Our `test_client` and `test_connection` modules contain pinpointed tests that aim to verify a single piece of our system, so we could move them into a unit package to signal that they are lightweight and can be run frequently. If I'm working on one of those classes, I'll know I'll be able to constantly run the tests related to it because they will be cheap.

So let's move them into a `tests/unit` package that we can run on demand:

```
└── tests
    ├── __init__.py
    ├── test_chat.py
    └── unit
        ├── __init__.py
        ├── test_client.py
        └── test_connection.py
```

Now we know that when working on specific parts of the system, we will be able to quickly verify them by running only the associated test unit:

```
$ python -m unittest discover tests/unit -v -k connection
test_client_connection (test_client.TestChatClient) ... ok
test_broadcast (test_connection.TestConnection) ... ok

----------------------------------------------------------------

Ran 2 tests in 0.006s

OK
```

The speed is such that test units could be run every time we save the file of the class that the tests aim to verify.

Our `test_chat.py` instead is very slow, but it verifies the system from client to server, and in real conditions, starts a real server over a real network. So let's make clear that its purpose is to verify the system **end to end (e2e)** by moving it into a `tests/e2e` package:

```
└── tests
    ├── __init__.py
    ├── e2e
    │   ├── __init__.py
    │   └── test_chat.py
    └── unit
        ├── __init__.py
        ├── test_client.py
        └── test_connection.py
```

There we will have the tests that run very slow and as we know this, we will probably want to run them only before making a new release of the software to confirm things work as expected on a real infrastructure:

```
$ python -m unittest discover tests/e2e -v
test_message_exchange (test_chat.TestChatAcceptance) ... ok
test_many_users (test_chat.TestChatMultiUser) ... ok
test_multiple_readers (test_chat.TestChatMultiUser) ... ok

----------------------------------------------------------------

Ran 3 tests in 3.568s

OK
```

OK, now we have tests that we can run when we modify a single component, and tests that we can run to confirm that the whole app runs correctly before making a new release.

But during development, how are we going to work on modifying existing features and add more? Trying to find each unit test we need to run to verify a feature is not very convenient, and also doesn't give us much confidence in the fact that those units will work well once they are set into the whole system.

On the other side, the e2e tests are too slow to base our development life cycle on them. If we add too many of them, we will have to wait for tests to run for more time than we actually spend coding. What we need is a set of tests that sit in the middle ground and verifies a function completely, but that are still able to run quickly enough that we can run them constantly during our development routine.

That goal is perfectly served by **functional tests**, a special set of integration tests that are expected to test a full feature, but are not required to reproduce the real conditions that the application will face out there in the wild. For example, the database can be fake, the parts of the system that are not involved in that feature could be disabled, or the networking could be replaced by the in-memory exchange of messages.

In our case, the slowness of our chat comes from the client-server communication, and the fact that in the test_connection.py module, we actually have a test_exchange_with_server test that tries a connection against a fake server. Thus we should get rid of the whole networking and server startup overhead like so:

```
def test_exchange_with_server(self):
    with unittest.mock.patch
                ("multiprocessing.managers.listener_client",
                    new={"pickle": (None, FakeServer())}):
        c1 = Connection(("localhost", 9090))
        c2 = Connection(("localhost", 9090))

        c1.broadcast("connected message")
        assert c2.get_messages()[-1] == "connected message"
```

In reality, that test doesn't suit the unit directory much, even if we might consider it a form of sociable unit test. Crossing the client-server boundary is usually a sign of a higher-level test, such as integration or e2e tests.

We could use that test as a foundation for our functional tests and move it to a functional/test_chat.py module that tests that our chat is able to send and receive messages using FakeServer. Instead of using Connection, we could rewrite the same test to actually use ChatClient (which uses Connection underneath) so that we can test that the functionality of exchanging messages with a server works as expected:

```
import unittest
from unittest import mock

from chat.client import ChatClient

from .fakeserver import FakeServer
```

```
class TestChatMessageExchange(unittest.TestCase):
    def setUp(self):
        self.srv = mock.patch("multiprocessing.managers.listener_client",
                                    new={"pickle": (None, FakeServer())})
        self.srv.start()

    def tearDown(self):
        self.srv.stop()
    def test_exchange_with_server(self):
        c1 = ChatClient("User1")
        c2 = ChatClient("User2")

        c1.send_message("connected message")
        assert c2.fetch_messages()[-1] == "User1: connected message"
```

Because we moved the `test_exchange_with_server` test out of our unit tests and into our functional tests, there is no more use for `FakeServer` in the unit tests, and it probably never really fit in there. So, we also moved the `FakeServer` class into a `fakeserver.py` module within the `functional` directory.

Then, our `TestChatMessageExchange` test case provides `setUp` and `tearDown` methods to enable a new `FakeServer` for each one of the tests within the case and disables it when the tests are complete. This allows us to write tests as if we were using a real server, without having to worry about the usage of a `FakeServer`.

Our functional tests are able to provide fairly good safety over the correctness of our features, but are going to run tens of times faster than the e2e tests. This is slower than the unit tests, but quick enough that we can frequently run them during our development routine:

```
$ python -m unittest discover -k functional -v
test_exchange_with_server
(tests.functional.test_chat.TestChatMessageExchange) ... ok

----------------------------------------------------------------------
Ran 1 test in 0.001s

OK
```

So we divided our test suite into three blocks: **e2e, functional**, and **unit**:

```
└── tests
    ├── __init__.py
    ├── e2e
    │   ├── __init__.py
    │   └── test_chat.py
```

```
├── functional
│   ├── __init__.py
│   ├── fakeserver.py
│   └── test_chat.py
└── unit
    ├── __init__.py
    ├── test_client.py
    └── test_connection.py
```

As software grows in complexity, you might feel the need to start having more kinds of integration tests, and as your code grows, you might want to explore introducing **narrow integration** tests (tests where you integrate only the few components you care about) instead of only having functional tests where the whole system is usually started. But this layout has proved to be a pretty good one for small/medium-sized projects over the years for me. The key is making sure that writing fast tests is convenient and that e2e tests can be easily rewritten as functional tests so that our expensive e2e tests remain in a minority.

Moving from e2e to functional

Take a look at our `TestChatMessageExchange.test_exchange_with_server` functional test that we wrote in the previous section:

```python
class TestChatMessageExchange(unittest.TestCase):
    ...

    def test_exchange_with_server(self):
        c1 = ChatClient("User1")
        c2 = ChatClient("User2")

        c1.send_message("connected message")
        assert c2.fetch_messages()[-1] == "User1: connected message"
```

It's probably easy to see that it looks a lot like our `TestChatAcceptance.test_message_exchange` e2e test:

```python
class TestChatAcceptance(unittest.TestCase):
    def test_message_exchange(self):
        with new_chat_server() as srv:
            user1 = ChatClient("John Doe")
            user2 = ChatClient("Harry Potter")

            user1.send_message("Hello World")
            messages = user2.fetch_messages()
            assert messages == ["John Doe: Hello World"]
```

The first one starts a new server, while the second one doesn't. But in the end, they both connect two users to a server, send a message from one user, and check that the other user received it.

The interesting difference, however, is that one takes nearly no time to run:

```
$ python -m unittest discover -k test_exchange_with_server -v
test_exchange_with_server
(tests.functional.test_chat.TestChatMessageExchange) ... ok

----------------------------------------------------------------------
Ran 1 test in 0.001s
```

While the other takes nearly a second to run:

```
$ python -m unittest discover -k test_message_exchange -v
test_message_exchange (tests.e2e.test_chat.TestChatAcceptance) ... ok

----------------------------------------------------------------------
Ran 1 test in 0.659s
```

As the two tests look very similar, could we maybe leverage the same approach to make a faster version of our other e2e tests so that we can still be sure that our chat is able to serve multiple users concurrently, without having to pay the cost of running slow e2e tests?

Yes, usually functional tests need to be able to exercise the whole system, so e2e tests can frequently be ported to be functional tests and benefit from their faster runtime. While we need a set of e2e tests to ensure that over a real network, things do work, we don't want to test every feature as an e2e test.

Most tests that start as e2e could be rewritten over time as functional tests to make our test suite able to keep up as our tests grow, but without sacrificing too much of the safety they provide, and while keeping our test suite fast.

So let's move the tests from the `TestChatMultiUser` e2e test case into the functional `TestChatMessageExchange` test case. The only thing we have to change in them is to remove the `with new_chat_server() as srv:` line as we no longer need to start a real server, but apart from that, they should be able to work as they are.

The `TestChatMessageExchange.setUp` method will take care of setting up a fake server for the tests – we just have to use the clients:

```
class TestChatMessageExchange(unittest.TestCase):
    ...

    def test_many_users(self):
```

```
        firstUser = ChatClient("John Doe")

        for uid in range(5):
            moreuser = ChatClient(f"User {uid}")
            moreuser.send_message("Hello!")

        messages = firstUser.fetch_messages()
        assert len(messages) == 5
    def test_multiple_readers(self):
        user1 = ChatClient("John Doe")
        user2 = ChatClient("User 2")
        user3 = ChatClient("User 3")

        user1.send_message("Hi all")
        user2.send_message("Hello World")
        user3.send_message("Hi")

        user1_messages = user1.fetch_messages()
        user2_messages = user2.fetch_messages()
        self.assertEqual(user1_messages, user2_messages)
```

Now that we have moved those tests to be functional tests, we are able to run a nearly complete check of our system in a few milliseconds by running the unit and functional tests:

```
$ python -m unittest discover -k functional -k unit
........
----------------------------------------------------------------------
Ran 8 tests in 0.007s

OK
```

Even running the whole test suite, including the e2e tests, now takes under a second, as we moved most of the expensive tests into lighter functional tests:

```
$ python -m unittest discover
.........
----------------------------------------------------------------------
Ran 9 tests in 0.661s

OK
```

Organizing the tests into the proper buckets is important to make sure our test suite is still able to run in a timeframe that can be helpful. If the test suite becomes too slow, we are just going to stop relying on it as working with it will become a frustrating experience.

That's why it's important to think about how to organize the test suite for your projects and keep in mind the various kinds of test suites that could exist and their goals.

Working with multiple suites

The separation of tests we did earlier in this chapter helped us realize that there can be multiple test suites inside our `tests` directory.

We can then point the `unittest` module to some specific directories using the `-k` option to run test units on every change, and functional tests when we think we have something that starts looking like a full feature. Thus, we will rely on e2e tests only when making new releases or merging pull requests to pass the last checkpoint.

There are a few kinds of test suites that are usually convenient to have in all our projects. The most common kinds of tests suites you will encounter in projects are likely the **compile suite**, **commit tests**, and **smoke tests**.

Compile suite

The compile suite is a set of tests that must run *very fast*. Historically, they were performed every time the code had to be recompiled. As that was a frequent action, the compile suite had to be very fast. They were usually static code analysis checks, and while Python doesn't have a proper compilation phase, it's still a good idea to have a compile suite that we can maybe run every time we modify a file.

A very good tool in the Python environment to implement those kinds of checks is the **prospector** project. Once we install prospector with `pip install prospector`, we will be able to check our code for any errors simply by running it inside our project directory:

```
$ prospector

Check Information
=================
 Started: 2020-06-02 15:22:53.756634
 Finished: 2020-06-02 15:22:55.614589
 Time Taken: 1.86 seconds
 Formatter: grouped
```

```
Profiles: default, no_doc_warnings, no_test_warnings, strictness_medium,
strictness_high, strictness_veryhigh, no_member_warnings
Strictness: None
Libraries Used:
Tools Run: dodgy, mccabe, pep8, profile-validator, pyflakes, pylint
Messages Found: 0
```

Our project doesn't currently have any errors, but suppose that in
the `ChatClient.send_message` method in `src/chat/client.py`, we mistype the
`sent_messages` variable, prospector would catch the error and notify us that we have a
bug in the code before we can run our full test suite:

```
$ prospector
Messages
========

src/chat/client.py
  Line: 23
    pylint: Unused variable 'sen_message' (col 8)
  Line: 24
    pylint: Undefined variable 'sent_message' (col 34)
  Line: 25
    pylint: Undefined variable 'sent_message' (col 15)
```

If your project relies on type hinting, prospector can also integrate **mypy** to verify the type
correctness of your software before you run the code for real, just to discover it won't work.

Commit tests

As the name suggests, commit tests are tests you run every time you commit a new change.
In our chat example project, the unit and functional tests would be our commit suite.

But as the project grows further and the functional tests start to get too slow, it's not
uncommon to see the functional tests become "push tests" that are only run before sharing
the code base with your colleagues, while the commit suite gets reduced to unit tests and
lighter forms of integration tests.

If you properly divided your test suite, which piece consists of your commit suite is usually just a matter of passing the proper -k option (one or multiple) to `unittest discover`:

```
$ python -m unittest discover -k unit -k functional
........
----------------------------------------------------------------------
Ran 8 tests in 0.007s

OK
```

Through the -k option we can select which parts of our test suite to run and thus limit the execution to only those tests that are fast enough to constitute our commit suite.

Smoke tests

Smoke tests are a set of tests used to identify whether we broke the system in an obvious way and thus let us know that it doesn't make sense to proceed with further testing.

Historically, it came from a time where test cases were manually verified, so before investing hours of human effort, a set of checks was performed to ensure that the system did work and thus it made sense to test it.

Nowadays, tests are far faster and cheaper as they are performed by machines, but it still makes sense to have a smoke test suite before running the more expensive tests. It's usually a good idea to select a subset of your e2e tests that constitute the smoke test suite, and run the complete e2e suite only if it passed the smoke tests.

Sometimes, smoke tests are a dedicated set of tests explicitly written for that purpose, but an alternative is to select a set of other tests that we know exercise the most meaningful parts of our system and "tag" them as smoke tests.

For example, if our e2e test suite had an extra `test_sending_message` test that verified that our `ChatClient` is able to connect to the server and send a message, that would be a fairly good candidate for our smoke test suite, as it doesn't make much sense to proceed with further e2e tests if we are not even able to send messages:

```
class TestChatAcceptance(unittest.TestCase):
    def test_message_exchange(self):
        ...

    def test_sending_message(self):
        with new_chat_server() as srv:
            user1 = ChatClient("User1")
            user1.send_message("Hello World")
```

More advanced testing frameworks frequently support the concept of "tagging" tests, so that we can run only those tests with a specific set of tags. But with unittest, it's still possible to build our smoke test suite simply by prefixing test names with the word smoke so that we can select them.

In this case, we would thus rename test_sending_message as test_smoke_sending_message to make it part of our smoke tests and we would be able to run our e2e tests as before, but also benefit from having a smoke test suite to run beforehand as our e2e tests grow further. So we will first have our smoke test, as follows:

```
$ python -m unittest discover -k smoke -v
test_smoke_sending_message (e2e.test_chat.TestChatAcceptance) ... ok

----------------------------------------------------------------------
Ran 1 test in 0.334s

OK
```

This is then followed by our e2e test:

```
$ python -m unittest discover -k e2e -v
test_message_exchange (e2e.test_chat.TestChatAcceptance) ... ok
test_smoke_sending_message (e2e.test_chat.TestChatAcceptance) ... ok

----------------------------------------------------------------------
Ran 2 tests in 0.957s

OK
```

As for the commit suite, we were able to rely on the -k option to only execute our smoke tests or all our e2e tests. Thus, we are able to select which kinds of tests we want to run.

Carrying out performance testing

Even though it's not related to verifying the correctness of software, a performance test suite is part of the testing strategy for many applications. Usually, they are expected to assess the performance of the software in terms of how fast it can do its job and how many concurrent users it can handle.

Due to their nature, performance tests are usually very expensive as they have to repeat an operation multiple times to get a benchmark that is able to provide a fairly stable report and absorb outliers that could have taken too long to run just because the system was busy doing something else.

For this reason, the performance test suite is usually only executed after all other suites are passed (also, it doesn't make much sense to assess how fast it can test the software when we haven't checked that it actually does the right thing).

For our chat example, we could write a benchmark suite that verifies how many messages per second we are able to handle:

1. To begin with, we don't want to put that into the middle of all the other tests, so we are going to put our benchmarks into a benchmarks directory, separate from the tests directory:

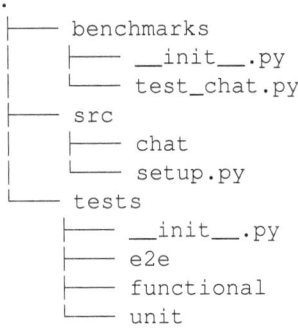

```
.
├── benchmarks
│   ├── __init__.py
│   └── test_chat.py
├── src
│   ├── chat
│   └── setup.py
└── tests
    ├── __init__.py
    ├── e2e
    ├── functional
    └── unit
```

2. test_chat.py can then contain the benchmarks we care about. In this case, we are going to create a benchmark to report how long it takes to send 10 messages:

```python
import unittest
import timeit

from chat.client import ChatClient
from chat.server import new_chat_server


class BenchmarkMixin:
    def bench(self, f, number):
        t = timeit.timeit(f, number=number)
        print(f"\n\ttime: {t:.2f}, iteration: {t/number:.2f}")


class BenchmarkChat(unittest.TestCase, BenchmarkMixin):
    def test_sending_messages(self):
        with new_chat_server() as srv:
            user1 = ChatClient("User1")

            self.bench(lambda: user1.send_message("Hello World"),
                       number=10)
```

BenchmarkMixin is a utility class that is going to provide the self.bench method we can use to report the execution time of our benchmarks. The real benchmark is provided by BenchmarkChat.test_sending_message, which is going to connect a client to a server and then repeat the user.send_message call 10 times.

3. Then we can run our benchmarks, pointing unittest to the benchmarks directory:

```
$ python -m unittest discover benchmarks -v
test_sending_messages (test_chat.BenchmarkChat) ...
        time: 2.31, iteration: 0.23
ok

_____
---
Ran 1 test in 2.406s
```

4. If we want to only run our tests instead, we could point the unittest module to the tests directory:

```
$ python -m unittest discover tests
..........
_____
---
Ran 10 tests in 1.013s
```

Running just python -m unittest discover will run both the benchmarks and tests, so make sure you point the discover process to the right directory when running your tests. An alternative is to name your benchmark files with a different prefix (bench_*.py instead of tests_*.py) and then use the -p option to specify the custom prefix when running your benchmarks. But in that case, it might not be immediately obvious how to run benchmarks for a new contributor to your project.

Our chat test suite is now fairly complete: it has e2e tests, functional tests, unit tests, smoke tests, and benchmarks. But we still have to remember to manually run all tests every time we do a change. Let's look at how we can tackle this.

Enabling continuous integration

Wouldn't it be convenient if someone else was in charge of running all our tests every time we made a change to our code base? This would mean that we couldn't forget to run some specific tests just because they were related to an area of the code that we were not directly touching.

That's exactly the goal of **Continuous Integration** (**CI**) environments. Every time we push our changes to the code repository, these environments will notice and rerun the tests, usually merging our changes with the changes from our colleagues to make sure they cope well together.

If you have a code repository on GitHub, using Travis as your CI is a fairly straightforward process. Suppose that I made an `amol-/travistest` GitHub project where I pushed the code base of our chat application; to enable Travis, the first thing that I have to do is to go to `https://travis-ci.com/` and log in with my GitHub credentials:

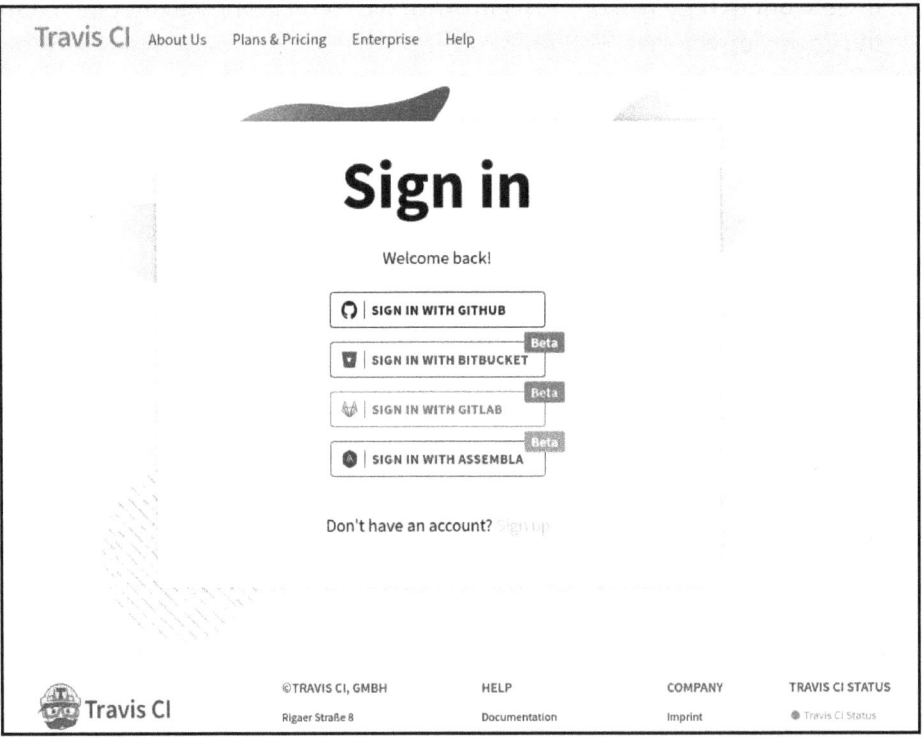

Figure 4.1 – Travis CI Sign in page

Once we are in, we must enable the integration with GitHub so that all our GitHub repositories become visible on Travis. We can do this by clicking on the **top-right profile icon** and then on the **Settings** option. That will show us a green **Activate** button that will allow us to enable Travis on our GitHub repositories:

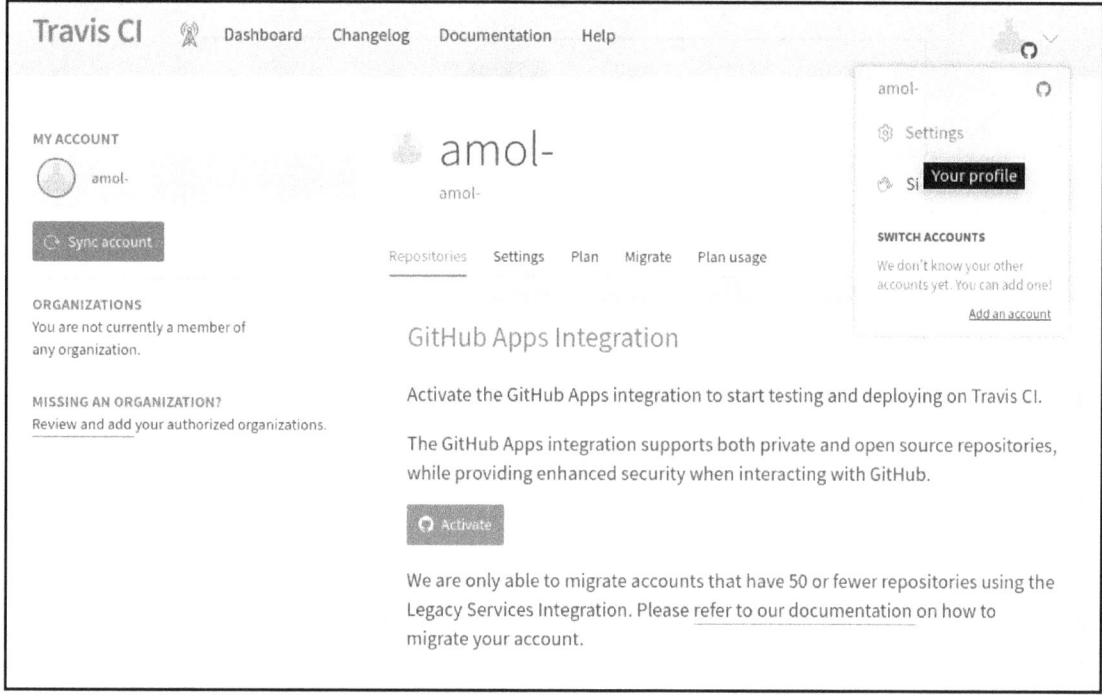

Figure 4.2 – Integrating with GitHub

Once we have enabled the Travis application on GitHub, we can go to `https://travis-ci.com/github/{YOUR_GITHUB_USER}/{GITHUB_PROJECT}` (which in my case is `https://travis-ci.com/github/amol-/travistest`) to confirm the repository is activated, but hasn't yet got any build:

Figure 4.3 – Confirming that the repository was activated

Travis will be monitoring your repository for changes. But it won't know how to run tests for your project. So even if we push changes to the source code, nothing will happen.

To tell Travis how to run our tests, we need to add to the repository a `.travis.yml` file with the following configuration:

```
language: python
os: linux
dist: xenial

python:
  - 3.7
  - &mainstream_python 3.8
  - nightly

install:
  - "pip install -e src"

script:
```

```
      - "python -m unittest discover tests -v"

  after_success:
      - "python -m unittest discover benchmarks -v"
```

This configuration is going to run our tests on Python 3.7, 3.8, and the current nightly build of Python (3.9 at the time of writing).

Before running the tests (the `install:` section), it will install the `chat` distribution from `src` to make the `chat` package available to the tests.

Then the tests will be performed as specified in the `script:` section and if they succeed, the benchmarks will be executed as stated in the `after_success:` section.

Once we push into the repository the `.travis.yml` file, Travis will see it and will start executing the tests as specified in the configuration file. If everything worked as expected, by refreshing the Travis project page, we should see a successful run of our tests on the three versions of Python:

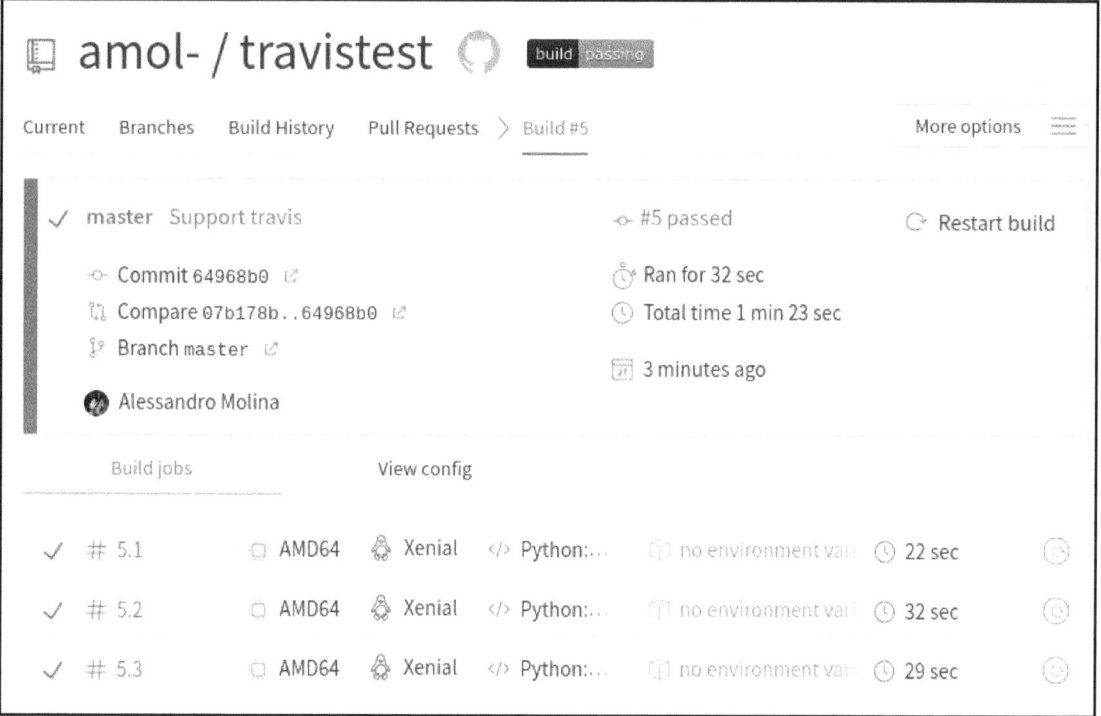

Figure 4.4 – Successful run on the three versions of Python

If you click on any of the jobs, it will show you what happened, confirming that both the tests and benchmarks were run:

```
160  $ git clone --depth=50 --branch=master                                git.checkout    0.65s
170
171  $ source ~/virtualenv/python3.7/bin/activate                                           0.01s
172  $ python --version
173  Python 3.7.1
174  $ pip --version
175  pip 19.0.3 from /home/travis/virtualenv/python3.7.1/lib/python3.7/site-packages/pip (python 3.7)
176  $ pip install -e src                                                    install         0.98s
181  $ python -m unittest discover tests -v                                                  0.99s
182  test_message_exchange (e2e.test_chat.TestChatAcceptance) ... ok
183  test_smoke_sending_message (e2e.test_chat.TestChatAcceptance) ... ok
184  test_exchange_with_server (functional.test_chat.TestChatMessageExchange) ... ok
185  test_many_users (functional.test_chat.TestChatMessageExchange) ... ok
186  test_multiple_readers (functional.test_chat.TestChatMessageExchange) ... ok
187  test_client_connection (unit.test_client.TestChatClient) ... ok
188  test_client_fetch_messages (unit.test_client.TestChatClient) ... ok
189  test_nickname (unit.test_client.TestChatClient) ... ok
190  test_send_message (unit.test_client.TestChatClient) ... ok
191  test_broadcast (unit.test_connection.TestConnection) ... ok
192
193  ----------------------------------------------------------------------
194  Ran 10 tests in 0.895s
195
196  OK
197  The command "python -m unittest discover tests -v" exited with 0.
198
199  $ python -m unittest discover benchmarks -v                             after_success   2.36s
200  test_sending_messages (test_chat.BenchmarkChat) ...
201          time: 2.24, iteration: 0.22
202  ok
203
```

Figure 4.5 – Checking the code base

Every time we make a change to our code base, Travis will rerun all tests, guaranteeing for us that we haven't broken anything and allowing us to see whether the performances became worse with the most recent changes.

Travis is not limited to performing a single thing such as running tests for your projects; it can actually perform multi-state pipelines that can be evolved to create releases of your packages or deploy them to a staging environment when the tests succeed. Just be aware that every build that you do will consume credits, and while you do have some available for free, you will have to switch to a paid plan if your CI needs grow beyond the amount covered by free credits.

Performance testing in the cloud

While our CI system does most of what we need, it's important to remember that cloud runners are not designed for benchmarking. So our performance test suite only becomes reliable when there are major slowdowns and over the course of multiple runs.

The two most common strategies when running performance tests in the cloud are as follows:

- To rerun the test suite multiple times and pick the fastest run, in order to absorb the temporary contention of resources in the cloud
- To record the metrics into a monitoring service such as Prometheus, from which it becomes possible to see the trend of the metrics over the course of multiple runs

Whichever direction you choose to go in, make sure you keep in mind that cloud services such as Travis can have random slowdowns due to the other requests they are serving, and thus it's usually better to make decisions over the course of multiple runs.

Summary

In this chapter, we saw how we can keep our test suite effective and comfortable as the complexity of our application and the size of our test suites grow. We saw how tests can be organized into different categories that could be run at different times, and also saw how we can have multiple different test suites in a single project, each serving its own purpose.

In general, over the previous four chapters, we learned how to structure our testing strategy and how testing can help us design robust applications. We also saw how Python has everything we need built in already through the `unittest` module.

But as our test suite grows and becomes bigger, there are utilities, patterns, and features that we would have to implement on our own in the `unittest` module. That's why, over the course of many years, many frameworks have been designed for testing by the Python community. In the next chapter, we are going to introduce **pytest**, the most widespread framework for testing Python applications.

Section 2: PyTest for Python Testing

In this section, we will learn how PyTest, the most widespread Python testing framework, can be applied to the concepts we learned in *Section 1, Software Testing and Test-Driven Development*, regarding plain Python. We will also learn how to set up fixtures and which plugins exist to make our lives easier when we're maintaining a test suite.

This section comprises the following chapters:

- Chapter 5, *Introduction to PyTest*
- Chapter 6, *Dynamic and Parametric Tests and Fixtures*
- Chapter 7, *Fitness Function with a Contact Book Application*
- Chapter 8, *PyTest Essential Plugins*
- Chapter 9, *Managing Test Environments with Tox*
- Chapter 10, *Testing Documentation and Property-Based Testing*

5
Introduction to PyTest

In the previous chapters, we saw how to approach test-driven development, how to create a test suite with the `unittest` module, and how to organize it as it grows. While `unittest` is a very good tool and is a reliable solution for most projects, it lacks some convenient features that are available in more advanced testing frameworks.

PyTest is currently the most widespread testing framework in the Python community, and it's mostly compatible with `unittest`. So it's easy to migrate from `unittest` to `pytest` if you feel the need for the convenience that `pytest` provides.

In this chapter, we will cover the following topics:

- Running tests with PyTest
- Writing PyTest fixtures
- Managing temporary data with `tmp_path`
- Testing I/O with capsys
- Running subsets of the test suite

Technical requirements

We need a working Python interpreter with the pytest framework installed. Pytest can be installed with the following:

```
$ pip install pytest
```

The examples have been written on Python 3.7 and pytest 5.4.3 but should work on most modern Python versions. You can find the code files present in this chapter on GitHub at https://github.com/PacktPublishing/Crafting-Test-Driven-Software-with-Python/tree/main/Chapter05.

Running tests with PyTest

PyTest is mostly compatible with the `unittest` module (apart from support for subtests). Test suites written with `unittest` can be directly run under `pytest` with no modification usually. For example, our chat application test suite can be directly run under `pytest` by simply invoking pytest within the project directory:

```
$ pytest -v
============ test session starts ============
platform linux -- Python 3.7.3, pytest-5.4.3, py-1.8.1, pluggy-0.13.1
cachedir: .pytest_cache
rootdir: /chatapp
collected 11 items

benchmarks/test_chat.py::BenchmarkChat::test_sending_messages PASSED [ 9%]
tests/e2e/test_chat.py::TestChatAcceptance::test_message_exchange PASSED [
18%]
tests/e2e/test_chat.py::TestChatAcceptance::test_smoke_sending_message
PASSED [ 27%]
tests/functional/test_chat.py::TestChatMessageExchange::test_exchange_with_
server PASSED [ 36%]
tests/functional/test_chat.py::TestChatMessageExchange::test_many_users
PASSED [ 45%]
tests/functional/test_chat.py::TestChatMessageExchange::test_multiple_reade
rs PASSED [ 54%]
tests/unit/test_client.py::TestChatClient::test_client_connection PASSED [
63%]
tests/unit/test_client.py::TestChatClient::test_client_fetch_messages
PASSED [ 72%]
tests/unit/test_client.py::TestChatClient::test_nickname PASSED [ 81%]
tests/unit/test_client.py::TestChatClient::test_send_message PASSED [ 90%]
tests/unit/test_connection.py::TestConnection::test_broadcast PASSED [100%]

============ 11 passed in 3.63s ============
```

The main difference is that `pytest` doesn't look for classes that inherit the `unittest.TestCase` class, but instead looks for anything that has **test** in the name, be it a module, a class, or a function. Anything named `[Tt]est*` is a test... but, if needed, it's possible to change the discovery rules by having `pytest.ini` inside the project directory.

This means that even a simple function can be a test as long as it's named `test_something`, and as it won't inherit from `TestCase`, there is no need to use the custom `self.assertEqual` and the related method to get meaningful information on failed assertions. Pytest will enhance the `assert` statement to report as much information as available on the asserted expression.

For example, we could create a very simple test suite that only has a `test_simple.py` module containing a `test_something` function. That would be all we need to start a test suite:

```
def test_something():
    a = 5
    b = 10
    assert a + b == 11
```

Now, if we run `pytest` inside the same directory, it will properly find and run our test, and the failed assertion will also give us hints on what went wrong by telling us that a + b is 15 and not 11:

```
$ pytest -v
======================= test session starts =======================
platform linux -- Python 3.7.3, pytest-5.4.3, py-1.8.1, pluggy-0.13.1
cachedir: .pytest_cache
rootdir: ~/HandsOnTestDrivenDevelopmentPython/05_pytest
collected 1 item

test_simple.py::test_something FAILED                        [100%]

============================= FAILURES =============================
_____ test_something _____

    def test_something():
        a = 5
        b = 10
>       assert a + b == 11
E       assert 15 == 11
E       +15
E       -11

test_simple.py:4: AssertionError
===================== short test summary info =====================
FAILED test_simple.py::test_something - assert 15 == 11
======================= 1 failed in 0.22s =======================
```

We can also add more complex tests that are implemented as classes collecting multiple tests, without having to inherit from the `TestCase` class as we did for `unittest` test suites:

```
class TestMultiple:
    def test_first(self):
        assert 5 == 5

    def test_second(self):
        assert 10 == 10
```

As for the previous case where we only had the `test_something` test function, if we run `pytest`, it will find all three tests and it will run them:

```
$ pytest -v
...
collected 3 items

test_simple.py::test_something FAILED [ 33%]
test_simple.py::TestMultiple::test_first PASSED [ 66%]
test_simple.py::TestMultiple::test_second PASSED [100%]
...
```

As we know that `test_something` always fails, we can select which tests to run by using the `-k` option, as we used to do for `unittest`. The option is, by the way, more powerful than the one provided by `unittest`.

For example, it is possible to provide the `-k` option to restrict the tests to a subset of them like we already used to do:

```
$ pytest -v -k first
...
collected 3 items / 2 deselected / 1 selected

test_simple.py::TestMultiple::test_first PASSED [100%]
...
```

It's also possible to use it to exclude some specific tests:

```
$ pytest -v -k "not something"
...
collected 3 items / 1 deselected / 2 selected

test_simple.py::TestMultiple::test_first PASSED [ 50%]
test_simple.py::TestMultiple::test_second PASSED [100%]
...
```

In the first case, we ran the `test_first` test, but in the second, we ran all tests except for `test_something`. So you could view `pytest` as `unittest` on steroids. It provides the same features you were used to with `unittest`, but frequently, they are enhanced to make them more powerful, flexible, or convenient.

If one had to choose between the two, it'd probably be a matter of preference. But it's not uncommon to see `unittest` used for projects that want to keep a more lightweight test suite that is kept stable over the course of the years (`unittest`, like most modules of the Python Standard Library guarantees very long-term compatibility) and `pytest` for projects that have more complex test suites or needs.

Writing PyTest fixtures

The primary difference between `unittest` and PyTest lies in how they handle fixtures. While `unittest` like fixtures (`setUp`, `tearDown`, `setupClass`, and so on) are still supported through the `TestCase` class when using `pytest`, `pytest` tries to provide further decoupling of tests from fixtures.

In `pytest`, a fixture can be declared using the `pytest.fixture` decorator. Any function decorated with the decorator becomes a fixture:

```
@pytest.fixture
def greetings():
    print("HELLO!")
    yield
    print("GOODBYE")
```

The code of the test is executed where we see the `yield` statement. `yield` in this context passes execution to the test itself. So this fixture would print `"HELLO"` before the test starts and then `"GOODBYE"` when the test finishes.

To then bind a fixture to a test, the `pytest.mark.usefixtures` decorator is used. So, for example, to use our new fixture with the existing `TestMultiple.test_second` test, we would have to decorate that test using the name of our new fixture:

```
class TestMultiple:
    def test_first(self):
        assert 5 == 5

    @pytest.mark.usefixtures("greetings")
    def test_second(self):
        assert 10 == 10
```

The name of a fixture is inherited by the name of the function that implements it, so by passing `"greetings"` to the `usefixtures` decorator, we end up using our own fixture:

```
$ pytest -v -k "usingfixtures and second" -s
...
collected 8 items / 7 deselected / 1 selected

test_usingfixtures.py::TestMultiple::test_second HELLO!
PASSED
GOODBYE
...
```

So, the part of the fixture before the `yield` statement replaces the `TestCase.setUp` method, while the part after `yield` replaces the `TestCase.tearDown` method.

If we want to use more than one fixture in a test, the `usefixtures` decorator allows us to pass multiple arguments, one for each fixture that we want to use.

If you are wondering about the `-s` option, that's another difference with `unittest`. By default, `pytest` captures all output that your code prints, while unittest, by default, didn't. The two work in a reverse way, so in the case of `pytest`, we need to explicitly disable output capturing to be able to see our prints.

Otherwise, outputs are only shown if the test fails. This has the benefit of keeping test run output cleaner, but can leave people puzzled the first time they see it.

Pytest fixtures can be declared in the same module that uses them, or inside a `conftest.py` module that will be inherited by all modules and packages in the same directory (or subdirectories).

Think of `conftest.py` as being a bit like the `__init__.py` of test packages; it allows us to customize tests' behavior for that package and even replace fixtures or plugins.

> While the `pytest` fixtures mechanism is very powerful, it's usually a bad idea to put fixtures too far away from what uses them.
>
> It will make it hard for tests reader to understand what's going on, so spreading tens of `conftest.py` files around the test suite is usually a good way to make life hard for anyone having to understand our test suite.
>
> As one of the primary goals of tests is to act as references of the software behavior, it's usually a good idea to keep them straightforward so that anyone approaching software for the first time can learn about the software without first having to spend days trying to understand how the test suite works and what it does.

Obviously, `pytest` fixtures are not limited to functions; they can also provide a replacement for `TestCase.setUpClass` and `TestCase.tearDownClass`. To do so, all we have to do is to declare a fixture that has `scope="class"` (`"function"`, `"module"`, `"package"`, and `"session"` scopes are available too to define the life cycle of a fixture):

```
@pytest.fixture(scope="class")
def provide_current_time(request):
    import datetime
```

```
request.cls.now = datetime.datetime.utcnow()

print("ENTER CLS")
yield
print("EXIT CLS")
```

In the previous fixture, we provide a `self.now` attribute in the class where the test lives, we print `"ENTER CLS"` before starting the tests for that class, and then we print `"EXIT CLS"` once all tests for that class have finished.

If we want to use the fixture, we just have to decorate a class with `mark.usefixtures` and declare we want it:

```
@pytest.mark.usefixtures("provide_current_time")
class TestMultiple:
    def test_first(self):
        print("RUNNING AT", self.now)
        assert 5 == 5

    @pytest.mark.usefixtures("greetings")
    def test_second(self):
        assert 10 == 10
```

Now, if we run our tests, we will get the messages from both the `provide_current_time` fixture and from the `greetings` one:

```
$ pytest -v -k "usingfixtures" -s
collected 8 items / 6 deselected / 2 selected

test_usingfixtures.py::TestMultiple::test_first
ENTER CLS
RUNNING AT 2020-06-17 22:28:23.489433
PASSED
test_usingfixtures.py::TestMultiple::test_second
HELLO!
PASSED
GOODBYE
EXIT CLS
```

You can also see that our test properly printed the `self.now` attribute, which was injected into the class by the fixture. The `request` argument to fixtures represents a request for that fixture from a test. It provides some convenient attributes, such as the class that requested the fixture (`cls`), the `instance` of that class that is being used to run the test, the `module` where the test is contained, the tests run `session`, and many more, allowing us not only to know the context of where our fixture is being used but also to modify those entities.

Apart from setting up tests, classes, and modules, there is usually a set of operations that we might want to do for the whole test suite; for example, configuring pieces of our software that we are going to need in all tests.

For that purpose, we can create a `conftest.py` file inside our test suite, and drop all those fixtures there. They just need to be declared with `scope="session"`, and the `autouse=True` option can automatically enable them for all our tests:

```python
import pytest

@pytest.fixture(scope="session", autouse=True)
def setupsuite():
    print("STARTING TESTS")
    yield
    print("FINISHED TESTS")
```

Now, running all our tests will be wrapped by the `setupsuite` fixture, which can take care of setting up and tearing down our test suite:

```
$ pytest -v -s
...
test_usingfixtures.py::TestMultiple::test_first
STARTING TESTS
ENTER CLS
RUNNING AT 2020-06-17 22:29:46.108487
PASSED
test_usingfixtures.py::TestMultiple::test_second
HELLO!
PASSED
GOODBYE
EXIT CLS
FINISHED TESTS
...
```

We can see from the output of the command that, according to our new fixture, the tests printed `"STARTING TESTS"` when they started and printed `"FINISHED TESTS"` at the end of the whole suite execution. This means that we can use session-wide fixtures to prepare and tear down resources or configurations that are necessary for the whole suite to run.

Using fixtures for dependency injection

Another good property of `pytest` fixtures is that they can also provide some kind of dependency injection management. For example, your software might use a remote random number generator. Whenever a new random number is needed, an HTTP request to a remote service is made that will return the number.

Inside our `conftest.py` file, we could provide a fixture that builds a fake random number generator that by default is going to generate random numbers (to test the software still works when the provided values are random) but without doing any remote calls to ensure the test suite is able to run quickly:

```
$ cat conftest.py

import pytest

@pytest.fixture
def random_number_generator():
    import random
    def _number_provider():
        return random.choice(range(10))
    yield _number_provider
```

Then, we could have any number of tests that use our random number generator (for the sake of simplicity, we are going to make a `test_randomness.py` file with a single test using it):

```
def test_something(random_number_generator):
    a = random_number_generator()
    b = 10
    assert a + b >= 10
```

If a test has an argument, `pytest` will automatically consider that dependency injection and will invoke the fixture with the same name of the argument to provide the object that should satisfy that dependency.

So, for our `test_something` function, the `random_number_generator` object is the one returned by our `random_number_generator` fixture, which returns numbers from 0 to 9.

As fixtures can be overridden inside modules or packages, if for some of our tests we wanted to replace the random number generator with a fairly predictable one (that always returns 1, all we would have to do is, again, declare a fixture with the same exact name inside the other module. Let's look at an example:

1. We would make a `test_fixturesinj.py` test module where we provide a new `random_number_generator` that is not random at all and we have a test that relies on that feature:

```
def test_something(random_number_generator):
    a = random_number_generator()
    b = 10
    assert a + b == 11


@pytest.fixture
def random_number_generator():
    def _number_provider():
        return 1
    yield _number_provider
```

2. If we run our two `test_something` tests, from the two modules, they will both pass, because one will be using a random number generator that builds random numbers, while the other will use one that always returns the number 1:

```
$ pytest -v -k "something and not simple"
...
collected 7 items / 5 deselected / 2 selected
test_fixturesinj.py::test_something PASSED [ 50%]
test_randomness.py::test_something PASSED [100%]
...
```

So we saw that `pytest` fixtures are much more flexible than `unittest` ones and that due to that greater decoupling and flexibility, great care has to be put into making sure it's clear which fixture implementations we end up using in our tests.

In the upcoming sections, we are going to look at some of the built-in fixtures that `pytest` provides and that are generally useful during the development of a test suite.

Managing temporary data with tmp_path

Many applications need to write data to disk. Surely we don't want data written during tests to interfere with the data we read/write during the real program execution. Data fixtures used in tests usually have to be predictable and we certainly don't want to corrupt real data when we run our tests.

So it's common for a test suite to have its own read/write path where all the data is written. If we decided the path beforehand, by the way, there would be the risk that different test runs would read previous data and thus might not spot bugs or might fail without a reason.

For this reason, one of the fixtures that `pytest` provides out of the box is `tmp_path`, which, when injected into a test, provides a temporary path that is always different on every test run. Also, it will take care of retaining the most recent temporary directories (for debugging purposes) while deleting the oldest ones:

```
def test_tmp(tmp_path):
    f = tmp_path / "file.txt"
    print("FILE: ", f)

    f.write_text("Hello World")

    fread = tmp_path / "file.txt"
    assert fread.read_text() == "Hello World"
```

The `test_tmp` test creates a `file.txt` file in the temporary directory and writes `"Hello World"` in it. Once the write is completed, it tries to access the same file again and confirm that the expected content was written.

The `tmp_path` argument will be injected by `pytest` itself and will point to a path made by `pytest` for that specific test run.

This can be seen by running our test with the `-s` option, which will make the `"FILE: ..."` string that we printed visible:

```
$ pytest test_tmppath.py -v -s
===== test session starts =====
...
collected 1 item
```

```
test_tmppath.py::test_tmp
FILE: /tmp/pytest-of-amol/pytest-3/test_tmp0/file.txt
PASSED

===== 1 passed in 0.03s =====
```

On every new run, the `pytest-3` directory will be increased, so the most recent directory will be from the most recent run and only the latest three directories will be kept.

Testing I/O with capsys

When we implemented the test suite for the TODO list application, we had to check that the output provided by the application was the expected one. That meant that we provided a fake implementation of the standard output, which allowed us to see what the application was going to write.

Suppose you have a very simple app that prints something when started:

```
def myapp():
    print("MyApp Started")
```

If we wanted to test that the app actually prints what we expect when started, we could use the `capsys` fixture to access the capture output from `sys.stdout` and `sys.stderr` of our application:

```
def test_capsys(capsys):
    myapp()

    out, err = capsys.readouterr()
    assert out == "MyApp Started\n"
```

The `test_capsys` test just starts the application (running `myapp`), then through `capsys.readouterr()` it retrieves the content of `sys.stdout` and `sys.stderr` snapshotted at that moment.

Once the standard output content is available, it can be compared to the expected one to confirm that the application actually printed what we wanted. If the application really printed "`MyApp Started`" as expected, running the test should pass and confirm that's the content of the standard output:

```
$ pytest test_capsys.py -v
===== test session starts =====
...
collected 1 item
```

```
test_capsys.py::test_capsys PASSED

===== 1 passed in 0.03s =====
```

The passing test run confirms that the `capsys` plugin worked correctly and our test was able to intercept the output sent by the function under test.

Running subsets of the testsuite

In the previous chapters, we saw how to divide our test suite into subsets that we can run on demand based on their purpose and cost. The way to do so involved dividing the tests by directory or by name, such that we could point the test runner to a specific directory or filter for test names with the `-k` option.

While those strategies are available on `pytest` too, `pytest` provides more ways to organize and divide tests; one of them being **markers**.

Instead of naming all our smoke tests `"test_smoke_something"`, for example, we could just name the test `"test_something"` and mark it as a smoke test. Or, we could mark slow tests, so that we can avoid running slow ones during the most frequent runs.

Marking a test is as easy as decorating it with `@pytest.mark.marker`, where `marker` is our custom label. For example, we could create two tests and use `@pytest.mark.first` to mark the first of the two tests:

```
import pytest


@pytest.mark.first
def test_one():
    assert True


def test_two():
    assert True
```

At this point, we could select which tests to run by using `pytest -m first` or `pytest -m "not first"`:

```
$ pytest test_markers.py -v
...
test_markers.py::test_one PASSED [ 50%]
test_markers.py::test_two PASSED [100%]
```

pytest test_markers.py -m "first" would run only the one marked with our custom marker:

```
$ pytest test_markers.py -v -m first
...
test_markers.py::test_one PASSED [100%]
```

This means that we can mark our tests in any way we want and run selected groups of tests independently from the directory where they sit or how they are named.

On some versions of pytest, you might get a warning when using custom markers:

```
Unknown pytest.mark.first - is this a typo?  You can register custom marks
to avoid this warning
```

This means that the marker is unknown to pytest and must be registered in the list of available markers to make the warning go away. The reason for this is to prevent typos that would slip by unnoticed if markers didn't have to be registered.

To make a marker available and make the warning disappear, the custom markers can be set in the pytest.ini configuration file for your test suite:

```
[pytest]
markers =
    first: mark a test as the first one written.
```

If the configuration file is properly recognized and we have no typos in the "first" marker, the previously mentioned warning will go away and we will be able to use the "first" marker freely.

Summary

In this chapter, we saw how pytest can provide more advanced features on top of the same functionalities we were already used to with unittest. We also saw how we can run our existing test suite with pytest and how we can evolve it to leverage some of built-in pytest features.

We've looked at some of the features that pytest provides out of the box, and in the next chapter, we will introduce more advanced pytest features, such as parametric tests and fixture generation.

6

Dynamic and Parametric Tests and Fixtures

In the previous chapter, we saw how `pytest` can be used to run our test suites, and how it provides some more advanced features that are unavailable in `unittest` by default. Python has seen multiple frameworks and libraries built on top of `unittest` to extend it with various features and utilities, but pytest has surely become the most widespread testing framework in the Python community. One of the reasons why pytest became so popular is its flexibility and support for dynamic behaviors. Apart from this, generating tests and fixtures dynamically or heavily changing test suite behavior are other features supported by pytest out of the box.

In this chapter, we are going to see how to configure a test suite and generate dynamic fixtures and dynamic or parametric tests. As your test suite grows, it will be important to be able to know which options PyTest provides to drive the test suite execution and how we can generate fixtures and tests dynamically instead of rewriting them over and over.

In this chapter, we will cover the following topics:

- Configuring the test suite
- Generating fixtures
- Generating tests with parametric tests

Technical requirements

We need a working Python interpreter with the pytest framework installed. Pytest can be installed using the following command:

```
$ pip install pytest
```

Though the examples have been written using Python 3.7 and pytest 5.4.3, they should work on most modern Python versions. You can find the code files used in this chapter on **GitHub** at `https://github.com/PacktPublishing/Crafting-Test-Driven-Software-with-Python/tree/main/Chapter06`

Configuring the test suite

In pytest, there are two primary configuration files that can be used to drive the behavior of our testing environment:

- `pytest.ini` takes care of configuring pytest itself, so the options we set there are mostly related to tweaking the behavior of the test runner and discovery. These options are usually available as command-line options too.
- `conftest.py` is aimed at configuring our tests and test suite, so it's the place where we can declare new fixtures, attach plugins, and change the way our tests should behave.

While pytest has grown over the years, with other ways being developed to configure the behavior of pytest itself or of the test suite, the two aforementioned ways are probably the most widespread.

For example, for a `fizzbuzz` project, if we have a test suite with the classical basic distinction between the source code, unit tests, and functional tests, then we could have a `pytest.ini` file within the project directory to drive how pytest should run:

```
.
├── pytest.ini
├── src
│   ├── fizzbuzz
│   │   ├── __init__.py
│   │   └── __main__.py
│   └── setup.py
└── tests
    ├── conftest.py
    ├── __init__.py
    ├── functional
    │   └── test_acceptance.py
    └── unit
        ├── test_checks.py
        └── test_output.py
```

The content of `pytest.ini` could contain any option that is also available via the command line, plus a bunch of extra options as described in the pytest reference for INI options.

For example, to run `pytest` in verbose mode, without capturing the output and by disabling deprecation warnings, we could create a `pytest.ini` file that adds the following related configuration options:

```
[pytest]
addopts = -v -s
filterwarnings =
    ignore::DeprecationWarning
```

In the same way, we have a `conftest.py` file in the `tests` directory. We already know from Chapter 5, *Introduction to PyTest*, that `conftest.py` is where we can declare our fixtures to make them available to the directory and all subdirectories. If set with `autouse=True`, the fixtures will also automatically apply to all tests in the same directory.

If we want to print every time we enter and exit a test, for example, we could add a fixture to our `conftest.py` file as shown here:

```
import pytest

@pytest.fixture(scope="function", autouse=True)
def enterexit():
    print("ENTER")
    yield
    print("EXIT")
```

As `conftest` is the entry point of our tests, the fixture would become available for all our tests, and as it is with `autouse=True`, all of them would start using it. Not only can we use fixtures that are declared in `conftest.py` itself, but we can also use fixtures that come from anything that was imported. We just have to declare the module as a plugin that has to be loaded when the tests start.

For example, we could have a `fizzbuzz.testing` package in our `fizzbuzz` project where `fizzbuzz.testing.fixtures` provides a set of convenience fixtures for anyone willing to test our simple app.

Similarly, we could have a `fizzbuzz.testing.fixtures.announce` fixture that announces every test being run:

```
import pytest

@pytest.fixture(scope="function", autouse=True)
```

```
def announce(request):
    print("RUNNING", request.function)
```

To use it, we just have to add our module to `pytest_plugins` in the `conftest.py` file as follows:

```
pytest_plugins = ["fizzbuzz.testing.fixtures"]
```

 Note that while `conftest.py` can be provided multiple times and will only apply to the package that contains it, `pytest_plugins` instead should only be declared in the root `conftest.py` file, as there is no way to enable/disable plugins on demand – they are always enabled for the whole test suite.

But adding fixtures is not all `conftest.py` can do. Pytest also provides a bunch of hooks that can be exposed from `conftest` (or from a plugin declared in `pytest_plugins`) that can be used to drive the behavior of the test suite.

The most obvious hooks are `pytest_runtest_setup`, which is called when preparing to execute a new test; `pytest_runtest_call`, called when executing a new test; and `pytest_runtest_teardown`, called when finalizing a test.

For example, our previous `announce` fixture can be rewritten using the `pytest_runtest_setup` hook as follows:

```
def pytest_runtest_setup(item):
    print("Hook announce", item)
```

Tons of additional hooks are available in pytest, such as a hook for parsing command-line options, a hook for generating test run reports, a hook for starting or finishing a whole test run, and so on.

 For a complete list of available hooks, refer to the pytest API reference at `https://docs.pytest.org/en/stable/reference.html#hooks`.

We have seen how to change the behavior of our test suite by using configuration options, `conftest`, and hooks, but pytest's flexibility doesn't stop there. Not only we can change the behavior of the test suite itself, but we can also change the behavior of the fixtures by generating those fixtures on demand.

Generating fixtures

Now that we know that `conftest.py` allows us to customize how our test suite should behave, the next step is to notice that pytest allows us to also change how our fixtures behave as well.

For example, the `fizzbuzz` program is expected to print `"fizz"` on every number divisible by 3, print `"buzz"` on every number divisible by 5, and print `"fizzbuzz"` on every number divisible by both.

To implement this, we could have `outfizz` and `outbuzz` functions that print `"fizz"` or `"buzz"` without a newline. This allows us to call each one of them to print `fizz` or `buzz` and to call both functions one after the other to print `fizzbuzz`.

To test this behavior, we could have a `tests/unit/test_output.py` module containing all the tests for our output utilities. For `outfizz` and `outbuzz`, we could write the tests as follows:

```
from fizzbuzz import outfizz, outbuzz, endnum

def test_outfizz(capsys):
    outfizz()

    out, _ = capsys.readouterr()
    assert out == "fizz"

def test_outbuzz(capsys):
    outbuzz()

    out, _ = capsys.readouterr()
    assert out == "buzz"
```

These are pretty simple tests that do the same thing, just over a different output. One invokes the `outfizz` function and checks whether it prints `"fizz"` and the other invokes the `outbuzz` function and checks whether it prints `"buzz"`.

We could think of having a form of dependency injection oriented toward the test itself, where the function to test is provided by a fixture. This would allow us to write the test once and provide two fixtures: one that injects `"fizz"` and one that injects `"buzz"`.

Going even further, we could even write the fixture only once, and dynamically generate it. Pytest allows us to parameterize fixtures. This means that out of a list of parameters, the fixture will run multiple times, once for each parameter. This will allow us to write a single test with a single fixture that injects the right function and the right expectation testing once for "fizz" and once for "buzz":

```
def test_output(expected_output, capsys):
    func, expected = expected_output

    func()

    out, _ = capsys.readouterr()
    assert out == expected
```

The test_output test relies on the expected_output fixture (which we will define shortly) and the capsys fixture that is provided by pytest itself.

The expected_output fixture will be expected to provide the function that generates the output we want to test (func) and the output we expect that function to print (expected).

Our expected_output fixture will get the function that generates the expected output from the fizzbuzz module through a getattr call that is meant to retrieve the outfizz and outbuzz functions:

```
import pytest

import fizzbuzz


@pytest.fixture(params=["fizz", "buzz"])
def expected_output(request):
    yield getattr(fizzbuzz, "out{}".format(request.param)), request.param
```

Thanks to @pytest.fixture(params=["fizz", "buzz"]), the expected_output fixture will be invoked by pytest twice, once for "fizz" and once for "buzz", leading to the test running twice, once for each parameter.

Through request.param the fixtures know which one of the two parameters is running, and using getattr, pytest retrieves from the fizzbuzz module the outfizz and outbuzz output generation functions and yields to the test the function to be tested and its associated expected output.

 When the parameters are not strings, you can also use the `ids` argument to provide a text description for them. This is so that when the tests runs, you know which parameter is being used.

We have seen how we can drive fixture generation from parameters and thus generate fixtures based on them, but pytest can go further and allow you to drive the fixture generation from command-line options.

For example, imagine that you want to be able to test two possible setups: one where the app prints lowercase strings, and another where it prints uppercase "FIZZ" and "BUZZ".

To do this, we could add to `conftest.py` a `pytest_addoption` hook to inject an extra `--upper` option in pytest:

```
def pytest_addoption(parser):
  parser.addoption(
      "--upper", action="store_true",
      help="test for uppercase behavior"
  )
```

When this option is set, the output functions will be tested for uppercase output.

We need to slightly modify our `expected_output` fixture to return the right uppercase string that we need to check against when the `--upper` option is provided:

```
@pytest.fixture(params=["fizz", "buzz"])
def expected_output(request):
    text = request.param
    if request.config.getoption("--upper"):
        text = text.upper()
    yield getattr(fizzbuzz, "out{}".format(request.param)), text
```

Now our tests, by default, will check against the lowercase output when run:

```
$ pytest -k output
tests/unit/test_output.py::test_output[fizz] PASSED
tests/unit/test_output.py::test_output[buzz] PASSED
```

But, if we provide `--upper`, we test against the uppercase output (which obviously makes our tests fail as the `outfizz` and `outbuzz` functions output text in lowercase):

```
$ pytest --upper -k output
tests/unit/test_output.py::test_output[fizz] FAILED
tests/unit/test_output.py::test_output[buzz] FAILED
========= FAILURES ==========
. . .
```

```
E AssertionError: assert 'fizz' == 'FIZZ'
...
E AssertionError: assert 'buzz' == 'BUZZ'
```

We have seen how to pass options to fixtures from parameters and the command line, but what if I want to configure fixtures from the tests themselves? That's possible thanks to markers. Using pytest.mark, we can add markers to our tests, and obviously, those markers can be read from the test suite and the fixtures. The most flexible thing about markers is that markers can have parameters too. So the markers can be used to set attributes for a specific test that will be visible to the fixture.

For example, we could have the tests be able to force lower/upper case configuration instead of relying on an external command-line option. The test could add a pytest.mark.textcase marker to flag whether it wants upper- or lowercase text from the fixture:

```
@pytest.mark.textcase("lower")
def test_lowercase_output(expected_output, capsys):
    func, expected = expected_output

    func()

    out, _ = capsys.readouterr()
    assert out == expected
```

Our test_lowercase_output is a perfect copy of test_output, apart from the added marker. The marker specifies that the test has to run with lowercase text even when the --upper option is provided.

To enable such a behavior, we have to modify our expected_output fixture to read the marker and its arguments. After reading the command-line options, we are going to retrieve the marker, get its first argument, and lower/upper the text based on it:

```
@pytest.fixture(params=["fizz", "buzz"])
def expected_output(request):
    text = request.param
    if request.config.getoption("--upper"):
        text = text.upper()
    textcasemarker = request.node.get_closest_marker("textcase")
    if textcasemarker:
        textcase, = textcasemarker.args
        if textcase == "upper":
            text = text.upper()
        elif textcase == "lower":
            text = text.lower()
        else:
```

```
            raise ValueError("Invalid Test Marker")
    yield getattr(fizzbuzz, "out{}".format(request.param)), text
```

Now if we run our test suite, while the `test_output` function will fail when we provide the `--upper` option (because the output functions provide only lowercase output), the `test_lowercase_output` test instead will always succeed because the fixture is configured by the test to only provide lowercase text:

```
$ pytest --upper -k output
tests/unit/test_output.py::test_output[fizz] FAILED
tests/unit/test_output.py::test_output[buzz] FAILED
tests/unit/test_output.py::test_lowercase_output[fizz] PASSED
tests/unit/test_output.py::test_lowercase_output[buzz] PASSED
```

We have seen how we can change the behavior of fixtures based on parameters and options that we provide, and many of those practices work out of the box for changing the behavior of the tests themselves. Just as fixtures have the `params` option, tests support the `@pytest.mark.parametrize` decorator, which allows generating tests based on parameters.

Generating tests with parametric tests

Sometimes you find yourself writing the same check over and over for multiple configurations. Instead, it would be convenient if we could write the test only once and provide the configurations in a declarative way.

That's exactly what `@pytest.mark.parametrize` allows us to do: to generate tests based on a template function and the various configurations that have to be provided.

For example, in our `fizzbuzz` software, we could have two `isbuzz` and `isfizz` checks that verify whether the provided number should lead us to print the `"buzz"` or `"fizz"` strings. Like always, we want to write a test that drives the implementation of those two little blocks of our software, and the tests might look like this:

```
def test_isfizz():
    assert isfizz(1) is False
    assert isfizz(3) is True
    assert isfizz(4) is False
    assert isfizz(6) is True


def test_isbuzz():
    assert isbuzz(1) is False
    assert isbuzz(3) is False
```

```
assert isbuzz(5) is True
assert isbuzz(6) is False
assert isbuzz(10) is True
```

The tests cover a few cases to make us feel confident that our implementation will be reliable, but it's very inconvenient to write the same check over and over for each possible number that we want to check.

That's where parameterizing the test comes into the picture. Instead of having the test_isfizz function be a long list of assertions, we could rewrite it to be a single assertion that gets rerun by pytest multiple times, once for each parameter it receives. The parameters could for example be the number to check with isfizz and the expected outcome, so that we can compare the outcome of invoking isfizz to the expected outcome:

```
@pytest.mark.parametrize("n,res", [
    (1, False),
    (3, True),
    (4, False),
    (6, True)
])
def test_isfizz(n, res):
    assert isfizz(n) is res
```

When we run the test suite, pytest will take care of generating all tests, one for each parameter, to guarantee we are checking all the conditions as we were doing before:

```
$ pytest -k checks
tests/unit/test_checks.py::test_isfizz[1-False] PASSED [ 20%]
tests/unit/test_checks.py::test_isfizz[3-True] PASSED [ 40%]
tests/unit/test_checks.py::test_isfizz[4-False] PASSED [ 60%]
tests/unit/test_checks.py::test_isfizz[6-True] PASSED [ 80%]
...
```

We can even go further and mix a fixture with a parameterized test and have the fixture generate one of the parameters. For the isfizz function, we explicitly provided the expected result; for the isbuzz test, we are going to have the fixture inject whether the number is divisible by 5 and thus whether it would print buzz or not.

To do so, we are going to provide a divisible_by5 fixture that does no more than to return whether the number is divisible by 5 or not:

```
@pytest.fixture(scope="function")
def divisible_by5(n):
    return n % 5 == 0
```

Then, we can have our parameterized test accept the parameter for the number, but use the fixture for the expected result, as shown in the following code:

```
@pytest.mark.parametrize("n", [1, 3, 5, 6, 10])
def test_isbuzz(n, divisible_by5):
    assert isbuzz(n) is divisible_by5
```

On each one of the generated tests, the number n will be provided to both the test and the fixture (by virtue of the shared argument name) and our test will be able to confirm that isbuzz returns True only for numbers divisible by 5:

```
$ pytest -k checks
...
tests/unit/test_checks.py::test_isbuzz[1] PASSED  [ 55%]
tests/unit/test_checks.py::test_isbuzz[3] PASSED  [ 66%]
tests/unit/test_checks.py::test_isbuzz[5] PASSED  [ 77%]
tests/unit/test_checks.py::test_isbuzz[6] PASSED  [ 88%]
tests/unit/test_checks.py::test_isbuzz[10] PASSED [100%]
```

It is also possible to provide arguments to the test through a fixture by using the indirect option of parametrize. In such a case, the parameter is provided to the fixture and then the fixture can decide what to do with it, whether to pass it to the test or change it. This allows us to replace test parameters, instead of injecting new ones as we did.

Summary

In this chapter, we saw why pytest is considered a very flexible and powerful framework for writing test suites. Its capabilities to automatically generate tests and fixtures on the fly and to change their behaviors through hooks and plugins are very helpful, allowing us to write smaller test suites that cover more cases.

The problem with those techniques is that they make it less clear what's being tested and how, so it's always a bad idea to abuse them. It's usually better to ensure that your test is easy to read and clear about what's going on. That way, it can act as a form of documentation on the behavior of the software and allow other team members to learn about a new feature by reading its test suite.

Only once all our test suites are written in a simple and easy-to-understand way can we focus on reducing the complexity of those suites by virtue of parameterization or dynamically generated behaviors. When dynamically generated behaviors get in the way of describing the behavior of software clearly, they can make the test suite unmaintainable and full of effects at a distance (due to the **Actions at a distance** anti-pattern) that make it hard to understand why a test fails or passes.

Now that we have seen how to write tests in the most powerful ways, in the next chapter we will focus on which tests to write. We are going to focus on getting the right fitness functions for our software to ensure we are actually testing what we care about.

Fitness Function with a Contact Book Application

7

We have already seen that in test-driven development, it is common to start development by designing and writing acceptance tests to define what the software should do and then dive into the details of how to do it with lower-level tests. That frequently is the foundation of **Acceptance Test-Driven Development (ATDD)**, but more generally, what we are trying to do is to define a **Fitness Function** for our whole software. A fitness function is a function that, given any kind of solution, tells us how good it is; the better the fitness function, the closer we get to the result.

Even though fitness functions are typically used in genetic programming to select the solutions that should be moved forward to the next iteration, we can see our acceptance tests as a big fitness function that takes the whole software as the input and gives us back a value of how good the software is.

All acceptance tests passed? This is 100% what it was meant to be, while only 50% of acceptance tests have been passed? That's half-broken... As far as our fitness function really describes what we wanted, it can save us from shipping the wrong application.

That's why acceptance tests are one of the most important pieces of our test suite and a test suite comprised solely of unit tests (or, more generally, technical tests) can't really guarantee that our software is aligned with what the business really wanted. Yes, it might do what the developer wanted, but not what the business wanted.

In this chapter, we will cover the following topics:

- Writing acceptance tests
- Using behavior-driven development
- Embracing specifications by example

Technical requirements

We need a working Python interpreter with the PyTest framework installed. For the behavior-driven development part, we are going to need the `pytest-bdd` plugin.

`pytest` and `pytest-bdd` can be installed using the following command:

```
$ pip install pytest pytest-bdd
```

The examples have been written on Python 3.7, pytest 6.0.2, and pytest-bdd 4.0.1, but should work on most modern Python versions. You can find the code files present in this chapter on GitHub at `https://github.com/PacktPublishing/Crafting-Test-Driven-Software-with-Python/tree/main/Chapter07`.

Writing acceptance tests

Our company has just released a new product; it's a mobile phone for extreme geeks that will only work through a UNIX shell. All the things our users want to do will be doable via the command line and we are tasked with writing the contact book application. Where do we start?

The usual way! First, we prepare our project skeleton. We are going to expose the contact book application as the `contacts` package, as shown here, and we are going to provide a main entry point. For now, we are going to invoke this with `python -m contacts`, but in the future, we will wrap this in a more convenient shortcut:

```
.
├── src
│   ├── contacts
│   │   ├── __init__.py
│   │   └── __main__.py
│   └── setup.py
└── tests
    ├── conftest.py
    ├── functional
    │   └── test_acceptance.py
    ├── __init__.py
    └── unit
```

For now, all our modules are empty, just placeholders are present, but the first thing we surely want to have is a location where we can place our acceptance tests. So, the `test_acceptance` module is born. Now, how do we populate it?

Our team applies an agile approach, so we have a bunch of user stories like stickers, with things such as *As a user, I want to have a command to add a new entry to my contact book, so that I can then call it without having to remember the number,* or *As a user, I want to have a way to remove contacts that I no longer require from my contact book, so that it doesn't get too hard to spot the contacts I care about.* While they might be enough for us to start imagining what the application is meant to do, they are far from being something that describes its behavior well enough to act as a fitness function.

So we pick one story, the one about being able to add new entries to the contact book application, and we start writing a set of acceptance tests for it that can describe its behavior in a more detailed way.

Writing the first test

So, we open the `tests/functional/test_acceptance.py` file and we write our first acceptance test. It has to run some kind of command line and then check that after a contact has been added to the list of contacts:

```
import contacts

class TestAddingEntries:
    def test_basic(self):
        app = contacts.Application()

        app.run("contacts add NAME 3345554433")

        assert app._contacts == [
            ("NAME", "3345554433")
        ]
```

We decide that `Application.run` will be the entry point of our application, so we just pass what the user wrote on the shell and it gets parsed and executed, and also decide that we are going to somehow store the contacts in a `_contacts` list. That's an implementation detail that we can change later on as we dive into the details of implementation, but for now it is enough to state that somehow we want to be able to see the contacts that we stored.

 Acceptance tests are meant to exercise the system from the user point of view and through the interfaces provided to the user. However, it is generally considered acceptable if the setup and assertion parts of the test access internals to properly prepare the test or verify its outcome. The important part is that the system is used from the user point of view.

Satisfied with the fact that we are now clear in our mind how we want the software to behave at a high level, we are eager to jump into the implementation. But remember that our acceptance tests are only as good because they are a proper fitness function.

Getting feedback from the product team

Our next step is to go back to someone from our product team and share our acceptance test with one of its members to see how good it is.

Luckily for us, our product team understands Python and they get back with a few points as feedback:

- People usually have a name and a surname and even middle names, so what happens when NAME contains spaces?
- We actually want to be able to store international numbers, so make sure you accept a "+" at the beginning of the phone numbers, but don't accept any random text. We don't want people wondering why their contacts don't work after they did a typo.

These points are all new acceptance criteria. Our software is good only if it's able to satisfy all these conditions. So, we go back to our editor and tweak our acceptance tests and we come back with the following:

```python
class TestAddingEntries:
    def test_basic(self):
        app = contacts.Application()

        app.run("contacts add NAME 3345554433")

        assert app._contacts == [
            ("NAME", "3345554433")
        ]

    def test_surnames(self):
        app = contacts.Application()

        app.run("contacts add Mario Mario 3345554433")
        app.run("contacts add Luigi Mario 3345554434")
        app.run("contacts add Princess Peach Toadstool 3339323323")

        assert app._contacts == [
            ("Mario Mario", "3345554433"),
            ("Luigi Mario", "3345554434"),
            ("Princess Peach Toadstool", "3339323323")
```

```
        ]

    def test_international_numbers(self):
        app = contacts.Application()

        app.run("contacts add NAME +393345554433")

        assert app._contacts == [
            ("NAME", "+393345554433")
        ]

    def test_invalid_strings(self):
        app = contacts.Application()

        app.run("contacts add NAME InvalidString")

        assert app._contacts == []
```

The test_surnames function now verifies that names with spaces work as expected, and that we also support multiple spaces for middle names and multiple surnames.

The test_international_numbers function now verifies that we support international phone numbers, while the test_invalid_strings function confirms that we don't save invalid numbers.

This should cover a fairly comprehensive description of all the behaviors our product team mentioned. Before declaring victory, we go back to our product people and review the acceptance tests with them.

One of the product team members points out that a key feature for them is that contacts have to be retained between two different runs of the application. As obvious as that might sound, our acceptance tests don't in any way exercise that condition, and so is an insufficient fitness function. A sub-optimal solution that lacks a major capability, such as loading back the contacts when you run the app the second time, would still pass all our tests and thus get the same grade as the optimal solution.

Back to our chair, we tweak our acceptance tests and add one more test that verifies that loading back contacts leads to the same exact list of contacts that we had before:

```
    class TestAddingEntries:
        ...

        def test_reload(self):
            app = contacts.Application()

            app.run("contacts add NAME 3345554433")
```

```
assert app._contacts == [
    ("NAME", "3345554433")
]

app._clear()
app.load()

assert app._contacts == [
    ("NAME", "3345554433")
]
```

The `test_reload` function largely behaves like our `test_basic` function up to the point where it clears any list of contacts currently loaded and then loads it again.

Note that we are not testing whether `Application._clear` does actually clear the list of contacts. In acceptance tests, we can take it for granted that the functions we invoke do what they are meant to do, but what we are interested in is testing the overall behavior and not how the implementation works.

The functional and units tests will verify for us whether the functions actually work as expected. From the acceptance tests, we can just use those functions, taking it for granted that they do work.

Now it is time for one more review with someone who has the goals of the software clear in mind and we can confirm that the acceptance tests now look good and cover what everyone wanted. The implementation can now start!

Making the test pass

Running our current test suite obviously fails because we have not yet implemented anything:

```
$ pytest -v
...
.../test_acceptance.py::TestAddingEntries::test_basic FAILED [ 25%]
.../test_acceptance.py::TestAddingEntries::test_surnames FAILED [ 50%]
.../test_acceptance.py::TestAddingEntries::test_international_numbers
FAILED [ 75%]
.../test_acceptance.py::TestAddingEntries::test_reload FAILED [100%]
...
    def test_basic(self):
>   app = contacts.Application()
E   AttributeError: module 'contacts' has no attribute 'Application'
```

The failed tests point us in the direction that we want to start by implementing the `Application` itself, so we can create our `tests/unit/test_application.py` file and we can start thinking about what the application is and what it should do.

As usual, we start by writing a bunch of functional and unit tests that drive our coding and testing strategies as we also grow the implementation. We then continue to add more unit and functional tests and implementation code until all our tests and acceptance tests pass:

```
$ pytest -v
functional/test_acceptance.py::TestAddingEntries::test_basic PASSED [ 5%]
functional/test_acceptance.py::TestAddingEntries::test_surnames PASSED [
10%]
functional/test_acceptance.py::TestAddingEntries::test_international_number
s PASSED [ 15%]
functional/test_acceptance.py::TestAddingEntries::test_invalid_strings
PASSED [ 20%]
functional/test_acceptance.py::TestAddingEntries::test_reload PASSED [ 25%]
unit/test_adding.py::TestAddContacts::test_basic PASSED [ 30%]
unit/test_adding.py::TestAddContacts::test_special PASSED [ 35%]
unit/test_adding.py::TestAddContacts::test_international PASSED [ 40%]
unit/test_adding.py::TestAddContacts::test_invalid PASSED [ 45%]
unit/test_adding.py::TestAddContacts::test_short PASSED [ 50%]
unit/test_adding.py::TestAddContacts::test_missing PASSED [ 55%]
unit/test_application.py::test_application PASSED [ 60%]
unit/test_application.py::test_clear PASSED [ 65%]
unit/test_application.py::TestRun::test_add PASSED [ 70%]
unit/test_application.py::TestRun::test_add_surname PASSED [ 75%]
unit/test_application.py::TestRun::test_empty PASSED [ 80%]
unit/test_application.py::TestRun::test_nocmd PASSED [ 85%]
unit/test_application.py::TestRun::test_invalid PASSED [ 90%]
unit/test_persistence.py::TestLoading::test_load PASSED [ 95%]
unit/test_persistence.py::TestSaving::test_save PASSED [100%]
```

For the sake of shortness, I won't report the implementation of all the tests that comprise our test suite. You can imagine that all unit tests had the purpose of checking the overall implementation and some specific corner cases.

The implementation of our `Application` class is fairly minimal. As we are going to evolve it with more tests in the next sections, we will make the implementation available here, allowing you to have a better understanding of the next sections:

```
class Application:
    PHONE_EXPR = re.compile('^[+]?[0-9]{3,}$')

    def __init__(self):
        self._clear()
```

```python
    def _clear(self):
        self._contacts = []

    def run(self, text):
        text = text.strip()
        _, cmd = text.split(maxsplit=1)
        cmd, args = cmd.split(maxsplit=1)

        if cmd == "add":
            name, num = args.rsplit(maxsplit=1)
            try:
                self.add(name, num)
            except ValueError as err:
                print(err)
                return
        else:
            raise ValueError(f"Invalid command: {cmd}")

    def save(self):
        with open("contacts.json", "w+") as f:
            json.dump({"_contacts": self._contacts}, f)

    def load(self):
        with open("contacts.json") as f:
            self._contacts = [
                tuple(t) for t in json.load(f)["_contacts"]
            ]

    def add(self, name, phonenum):
        if not isinstance(phonenum, str):
            raise ValueError("A valid phone number is required")

        if not self.PHONE_EXPR.match(phonenum):
            raise ValueError(f"Invalid phone number: {phonenum}")

        self._contacts.append((name, phonenum))
        self.save()
```

The `Application.add` method is the one that is explicitly in charge of adding new contacts to the contacts list, and it's what most of our tests rely on when they want to add new contacts. The `Application.save` and `Application.load` methods are now in charge of adding a persistence layer to the application. For the sake of simplicity, we just store the contacts in a JSON file (in the real world, you might want to change where the contacts are saved or make it configurable, but for our example, they will just be saved locally where the command is invoked from).

Finally, `Application.run` is the user interface to our software. Given any command in the form `"executablename command arguments"`, it parses it and executes the correct command. Currently, only `add` is implemented, but in the following sections, we will implement the `del` and `ls` commands, too.

Now we know that acceptance tests are vital in the feedback cycle in the case of people that understand the goals of the software well. Next, we need to focus on how to improve that communication cycle. In this example we were lucky that our counterpart understood Python, but what if they didn't? It's probably common that the people who understand the business well don't know a thing about programming, and so we need a better way to set up our communication than using Python.

Using behavior-driven development

For the first phase of our contact book application, we took it for granted that the people we had to speak with understood the Python language well enough that we could share with them our acceptance tests for review and confirm that we were going in the right direction.

While it's getting more and more common that people involved in product definition have at least an entry-level knowledge of programming, we can't take it for granted that every stakeholder we need to enter into discussions with knows Python.

So how can we keep the same kind of feedback loop and apply the strategy of reviewing all our acceptance tests with other stakeholders without involving Python?

That's what **Behavior-Driven Development (BDD)** tries to solve. BDD takes some concepts from **Test-Driven Development (TDD)** and **Domain-Driven Design (DDD)** to allow all stakeholders to speak the same language as the technical team.

In the end, BDD tries to mediate between the two worlds. The language becomes English, instead of Python, but a more structured form of English for which, in the end, a parser can be written and developers embrace the business glossary (no more classes named `User` and `PayingUser` if the business calls them `Lead` and `Customer`) so that the tests that the developers write make sense for all other stakeholders, too.

This is usually achieved by defining the tests in a language that is commonly named **Gherkin** (even though BDD doesn't strictly mandate Gherkin usage) and, luckily for us, the `pytest-bdd` plugin allows us to extend our test suite with tests written in a subset of the Gherkin language that coexists very well with all the other `pytest` plugins or features we might be using.

Our application is able to add contacts, but it still doesn't allow us to delete or list them, so it's not very useful. So, the next step is to implement a *delete contacts* feature, and we decide to do so by using BDD.

To get started using BDD, we will create a new `tests/acceptance` directory, where we are going to put all the acceptance tests for our features. Thus, the final layout of our test suite will appear as follows:

```
└── tests
    ├── __init__.py
    ├── conftest.py
    ├── acceptance
    │   └── ...
    ├── functional
    │   └── test_acceptance.py
    └── unit
        └── ...
```

Then we can create a `tests/acceptance/delete_contact.feature` file that will contain all our acceptance scenarios for the deleting contacts feature.

Defining a feature file

We start the file by making it clear that it covers the deletion of contacts:

```
Feature: Deleting Contacts
    Contacts added to our contact book can be removed.
```

Now that we have a location where all our testing scenarios can reside, we can try to add the first one. The basics of the Gherkin language are fairly easy to grasp. In the end, it is meant to be readable by everyone without having to study a programming language. So, the core words are the `Given`, `When`, `Then`, and `And` keywords that start every step in our scenarios, and to start a new scenario we just use `Scenario`.

Declaring the scenario

After the feature definition, we declare that our first scenario tries to delete a basic contact and see that things work as expected:

```
Scenario: Removing a Basic Contact
    Given I have a contact book
    And I have a "John" contact
    When I run the "contacts del John" command
    Then My contacts list is now empty
```

The scenario is written in fairly plain English so that we can review with other stakeholders without having to understand software development or programming languages. Once we agree that it represents correctly what we expect from the software, then we can turn it to code by using `pytest-bdd`.

`pytest-bdd` is based on PyTest itself, so each scenario is exposed as a test function. To signal that it's also a scenario, we add the `@scenario` decorator and point it to the feature file.

In our `tests/functional/test_acceptance.py` file, we are going to add a test for our `Removing a Basic Contact` scenario that we described in the `tests/acceptance/delete_contact.feature` file using the Gherkin language:

```python
from pytest_bdd import scenario

@scenario("../acceptance/delete_contact.feature",
            "Removing a Basic Contact")
def test_deleting_contacts():
    pass
```

Unlike the standard PyTest test, a scenario test is usually an empty function. We can perform additional testing within the function, but any code that we add will run after the scenario has been completed.

The test itself is loaded from the feature file looking for a scenario that has the same name as the one we provided in the decorator (in this case,`"Removing a Basic Contact"`). Then, the scenario text is parsed and each step defined in it is executed one after the other.

But how does PyTest know what to do in order to perform the steps? Our scenario starts by ensuring that we have a contact book with one contact inside named `"John"`:

```
Given I have a contact book
And I have a "John" contact
```

How does `pytest-bdd` even know what a contact book is and how to add a contact to it? Well, it doesn't.

Running the scenario test

If we try to run our scenario test at this point, as shown here, PyTest will just report errors complaining that it doesn't know what to do with the first step that it encounters:

```
$ pytest -v -k deleting
.../test_acceptance.py::test_deleting_contacts FAILED [100%]
...
StepDefinitionNotFoundError: Step definition is not found: Given "I have a
contact book". Line 5 in scenario "Removing a Basic Contact" in the feature
"/tests/acceptance/delete_contact.feature
```

We have to tell `pytest-bdd` what to do when it faces a step. So, in our `test_acceptance.py`, we must provide the code that has to be executed when a step is met and link it to the step using the `@given` decorator, as shown in the following code block:

```
from pytest_bdd import scenario, given

@given("I have a contact book", target_fixture="contactbook")
def contactbook():
    return contacts.Application()
```

Every time `pytest-bdd` finds "I have a contact book" as a step in a scenario, it will invoke our `contactbook` function. The `contactbook` function doesn't just tell PyTest how to run the step, but thanks to the `target_fixture="contactbook"` argument of `@given`, it's also a fixture that provides the `contactbook` dependency every time it is requested by another step. Any other step of the scenario that requires a `contactbook` can just refer to it and they will get back the `Application` that was created by the "I have a contact book" step.

Further setup with the And step

For now, the contact book is totally empty, but we know that the next step is to add a contact named "John" to it:

```
And I have a "John" contact
```

In this case, we have to tell `pytest-bdd` how to add contacts to a contact book and that `John` is the name of the contact that we want to add. This can be done by using another `@given` decorator (`And` is an alias for the previous keyword we just used):

```
from pytest_bdd import parsers

@given(parsers.parse("I have a \"{contactname}\" contact"))
def have_a_contact(contactbook, contactname):
    contactbook.add(contactname, "000")
```

We are also relying on `parsers.parse` provided by `pytest-bdd` to let the step definition know that `"John"` is not part of the step itself, but that it's actually a variable. It can be `John`, it can be `Jane`, it can be any name, and they will all go to this same step. The name of the contact will be extracted from the step and will be passed as an argument to the function in charge of executing the step.

Our function then only has to take that name and add it to the contact book. But where does the contact book come from? When we declared the `"I have a contact book"` step, we said that steps can also be PyTest fixtures, so when our `have_a_contact` function finds the need for a `contactbook` argument, the PyTest dependency injection will resolve it for us by providing what the `contactbook` fixture that was associated with the just executed `"I have a contact book"` step returned.

Hence, to the `contactbook` provided by the fixture, we invoke the `add` method, passing the `contactname` provided by the parser. In this scenario, we don't care for phone numbers, so the contact is always added with `"000"` as its phone number.

Performing actions with the When step

Moving forward, our next step in the scenario is a `When` step. These are no longer steps associated with preparation for testing; they are intended to perform the actions we want to perform (remember the `Arrange`, `Act`, and `Assert` pattern? Well, we could consider the `Given`, `When`, and `Then` steps as the three BDD counterparts to the pattern):

```
When I run the "contacts del John" command
```

Just like the previous two steps, we are going to link a function in charge of executing the code that has to happen when this step is found, in this case using the `pytest_bdd.when` decorator:

```
from pytest_bdd import when

@when(parsers.parse("I run the \"{command}\" command"))
```

```
def runcommand(contactbook, command):
    contactbook.run(command)
```

As we did for the `have_a_contact` function, `runcommand` needs the `contactbook` against which it has to run the command and can rely on `parsers.parse` to retrieve the command that has to be executed on the contact book.

It's not uncommon that the `Given` and `When` steps can be reused across multiple scenarios. Making those steps parametric using parsers allows their implementation functions to be reused more frequently. In this case, we will be able to reuse the same step definition independently from the command we want to run, thus allowing us to implement scenarios for other features too, and not just for deleting contacts.

In this case, as our step was `I run the "contacts del John"`command, the function will run the `contacts del John` command as this is the one we have provided in the step.

Our final step in the scenario is the one meant to verify that the contact was actually deleted as we expect once the command is performed:

```
Then My contacts list is now empty
```

Then, steps usually translate into the final assertion phase of our tests, so we are going to verify that the contact book is really empty.

Assessing conditions with the Then step

In this case, there is no need to parse anything, but our function will still need the `contactbook` for which it has to verify that it is actually empty:

```
from pytest_bdd import then

@then("My contacts book is now empty")
def emptylist(contactbook):
    assert contactbook._contacts == []
```

Now that we have provided the entry point for our scenario and the implementation of all its steps, we can finally retry running our tests to confirm that the scenario actually gets executed:

```
$ pytest -v -k deleting
.../test_acceptance.py::test_deleting_contacts FAILED [100%]
...
```

```
E ValueError: Invalid command: del
src/contacts/__init__.py:27: ValueError
```

Our scenario steps were all properly executed, but, as expected, our test has not passed. It choked on the When I run the "contacts del John"command step because our contacts application doesn't yet recognize the del command.

Making the scenario pass

So, our next steps will involve diving into the functional and unit tests that we need to define how the del command has to behave while providing an implementation for it.

As that's a part that we already know from the previous chapters of the book, we are going to provide the final resulting implementation here directly:

```python
class Application:
    ...

    def run(self, text):
        text = text.strip()
        _, cmd = text.split(maxsplit=1)
        cmd, args = cmd.split(maxsplit=1)

        if cmd == "add":
            name, num = args.rsplit(maxsplit=1)
            try:
                self.add(name, num)
            except ValueError as err:
                print(err)
                return
        elif cmd == "del":
            self.delete(args)
        else:
            raise ValueError(f"Invalid command: {cmd}")

    ...
    def delete(self, name):
        self._contacts = [
            c for c in self._contacts if c[0] != name
        ]
        self.save()
```

Now that our implementation is in place and the `"del"` command is dispatched to the `Application.delete` function, it will remove anyone matching the provided name from the list of contacts. We can check that our acceptance test passes and that our contacts book application is actually doing what we meant it to do:

```
$ pytest -v -k deleting
.../test_acceptance.py::test_deleting_contacts    PASSED [100%]
...
```

Our scenario was executed and our implementation satisfied it. The steps were executed by the functions we provided in our `test_acceptance.py` file:

```python
@scenario("../acceptance/delete_contact.feature",
          "Removing a Basic Contact")
def test_deleting_contacts():
    pass

@given("I have a contact book", target_fixture="contactbook")
def contactbook():
    return contacts.Application()

@given(parsers.parse("I have a \"{contactname}\" contact"))
def have_a_contact(contactbook, contactname):
    contactbook.add(contactname, "000")

@when(parsers.parse("I run the \"{command}\" command"))
def runcommand(contactbook, command):
    contactbook.run(command)

@then("My contacts book is now empty")
def emptylist(contactbook):
    assert contactbook._contacts == []
```

The problem with this approach is that if we have multiple scenarios, then it can tend to get confusing. It's already hard to spot out of the box the order of execution of this code, or the relations between the functions. We would have to constantly jump back and forth to the `.feature` file in order to understand what's going on.

This is especially the case if we have multiple different scenarios from unrelated features that can become hard to navigate, making it difficult to even distinguish between scenarios that are related to the same feature.

For this reason, people tend to split the features into multiple Python modules. Each Python module will contain the functions implementing the scenarios and steps that are only related to that, usually leading to a layout that is similar to `tests/acceptance/deleting_contacts.py`, `tests/acceptance/adding_contacts.py`, and so on.

Now that we know how to write acceptance tests in a more shareable way, we are going to lower the barrier of how easy they are to understand and verify for a human by introducing specification by example, a practice that tries to ensure that what the software has to do is not only expressed and verified, but that it is also expressed in a way that is less subject to misunderstandings.

With BDD, we might all agree on what's written in the acceptance tests and say that it expresses perfectly the specifications of our software, but the translation phase from the Gherkin syntax to code based tests can lead to misunderstandings. Specifications by example try to solve these kinds of issues by relying on clear examples that should be hard to misunderstand and by providing multiple examples for each scenario to further reduce doubts.

Embracing specifications by example

A common problem with acceptance tests is that it takes some effort to understand what's going on. If you are not already familiar with the domain, it can be easy to misunderstand them, thus leading to the wrong checks being performed even if everyone that reviewed it agreed with the original acceptance tests.

For example, if I read an acceptance test such as the following:

```
Given a first number 2
And a second number 3
When I run the software
Then I get 3 as the output
```

I might be tempted to understand it as, *Oh, ok! The test is meant to verify that given two numbers, we print the highest one.*

But that might not be the requirement; the requirement might actually be, *Given two numbers, print the lowest one plus one.* How can I understand which one that test was actually meant to verify?

The answer is to provide more examples. The more examples we provide for our tests, the easier it is to understand them.

Examples are provided in a table-like format, where columns are meant to show the data involved in our examples and the resulting outcomes. In general, we can say that the columns should describe the state of the system for that example:

```
Number1 | Number2 | Result
   2    |    3    |   3
```

If, by having only 2 and 3 as numbers and 3 as the result, both understandings of the test would be acceptable, the moment I expand my examples with one more, it becomes immediately obvious which one of the two I meant.

So we can add one more row to our examples table to add an example that further reduces the uncertainty regarding what the expected behavior is:

```
Number1 | Number2 | Result
   2    |    3    |   3
   5    |    7    |   6
```

The second example makes it possible to understand that we are not printing the highest of the two numbers, but that we are actually printing the lowest plus one.

What if I have further doubts? Maybe it's not the lowest plus one; maybe it's the first of the two numbers plus one!

```
Number1 | Number2 | Result
   2    |    3    |   3
   5    |    7    |   6
   8    |    4    |   5
```

With the third example, we made it clear that we actually want the lowest of the two numbers and not the first one. Just add more examples until the reading of the test becomes fairly obvious for every reader.

That's the core idea behind specification by example: the behavior of a software can be described by providing enough examples that make it obvious to see what's going on.

Instead of having to write tens of pages trying to explain what's happening, given enough examples, which can be automatically verified, the reader can easily see what's going on.

Generally, there are many benefits to this approach, including the following:

- We don't have *the specification* and *the test*: the specifications are testable by definition.
- Tests that were easy to misunderstand can easily be made more obvious by adding more examples, which are cheaper to add than more tests.

- You can't change the behavior of the software without updating the specifications. The specifications are the examples used to verify the software; if they don't verify, then the updated tests would not pass.

As the specifications are meant to be human-readable, the Gherkin language is a good foundation for writing the specifications themselves making sure that they can be verified. We just need to add a section where we provide a list of all the possible examples for a scenario.

For example, we might write the final feature of our software: *Listing the contacts using this model*. To do so, let's write a scenario with two examples of possible contact lists to print:

```
Feature: Listing Contacts
    Contacts added to our contact book can be listed back.

Scenario: Listing Added Contacts
    Given I have a contact book
    And I have a first <first> contact
    And I have a second <second> contact
    When I run the "contacts ls" command
    Then the output contains <listed_contacts> contacts

    Examples:
    | first | second | listed_contacts |
    | Mario | Luigi  | Mario,Luigi     |
    | John  | Jane   | John,Jane       |
```

Compared to the scenarios we wrote before, the main difference is that we used some placeholders contained within angular brackets (`<first>`, `<second>`, and `<listed_contacts>`), and then we have a list of examples at the end of the scenario.

This whole feature description with its examples becomes our specification and sole document that we discuss with all stakeholders. If we have doubts, we add more examples and scenarios to the feature until it becomes obvious to everyone how the software should behave.

We save our feature description as `"tests/acceptance/list_contacts.feature"` and, as we did for the previous cases, we start by adding a test for our scenario so that PyTest knows that we have one more test to run:

```
@scenario("../acceptance/list_contacts.feature",
            "Listing Added Contacts")
def test_listing_added_contacts(capsys):
    pass
```

As we have to check the output of the command (which will print the contacts), this time, our test explicitly mentions the `capsys` fixture, so that output starts to be captured when the test is run.

The first step of our scenario is "`Given I have a contact book`", which we had already implemented for our previous contacts deletion test, so in this case we have nothing to do. `pytest-bdd` will reuse the same test implementation as the step is the same.

Going further, we have two steps in charge of adding the two contacts from the examples into our contact list:

```
And I have a first <first> contact
And I have a second <second> contact
```

These translate into two new steps, and both of these are in charge of adding one contact to the contact book, as shown in the following code block:

```
@given("I have a first <first> contact")
def have_a_first_contact(contactbook, first):
    contactbook.add(first, "000")
    return first
@given("I have a second <second> contact")
def have_a_second_contact(contactbook, second):
    contactbook.add(second, "000")
    return second
```

As the two tests have the same exact implementation, you might be wondering why we made two different `Given` steps instead of a single one with a parser.

The reason is because `Given` steps, in BDD, are meant to represent data that is needed to perform the test. They state what you have in a way that should make it possible to look up any of the given things explicitly. If, in any other step, we want to know what's the name of the first person that was added to the contact book, that step would only have to refer to the given test by the name of the function, and the given step would behave as a fixture providing that specific entity.

To make it easier to understand, if we want to get back the name of the first contact added to the contact book, we just have to add a `have_a_first_contact` argument to the function implementing the step that needs that name. As the `have_a_first_contact` function returns a value, that value would be associated with any `have_a_first_contact` argument name in any other step.

In the same way, if we want to refer to the second person in our contact book, we just have to require the `have_a_second_contact` argument.

If, instead of having those two separate `Given` steps, we had a single `have_a_contact` step that used a parser, and we used it twice to add two contacts, which one of the two would the `have_a_contact` argument refer to? It would be ambiguous, and that's why `pytest-bdd` prevents reuse of the same `Given` step twice in the same scenario. Each `Given` step must be unique so that the data it provides is uniquely identifiable by the step name.

 The same doesn't apply to other kinds of steps. For example, it's perfectly possible to reuse the same `When` step multiple times in a scenario. That's because `When` steps are not meant to represent data and so have no need to be uniquely identifiable.

Now that we have our `Given` steps in place, the next step is the `When` step, which is meant to run the command that lists our contacts:

```
When I run the "contacts ls" command
```

This again is a step that we already implemented in our previous delete contact scenario. In the scenario, the `When` step we implemented there accepted a command to run as an argument, and so it's able to run any command. `pytest-bdd` will be able to reuse it, and hence we don't have to implement anything.

The final step is the one meant to verify that the command actually did what we expect, the `Then` step:

```
Then the output contains <listed_contacts> contacts
```

This step will have to check the output provided by the command and ensure that the contacts we wrote in our example actually exist in the output:

```python
@then("the output contains <listed_contacts> contacts")
def outputcontains(listed_contacts, capsys):
    expected_list = "".join([
        f"{c} 000\n" for c in listed_contacts.split(",")
    ])
    out, _ = capsys.readouterr()
    assert expected_list == out
```

We already know that we need `capsys` to be able to read the output of a program being tested. Apart from `capsys`, our step also requires the list of contacts that it has to check. Those are coming from the `Examples` section in the scenario.

In the `Examples` entry, `listed_contacts` were provided as comma-separated ("`Mario,Luigi`"), so the first thing we do is to split them by the comma so that we can get back all the contacts. Then, as our program is going to print them in separate lines with their phone numbers, we append the phone number at the end of the line (which is hardcoded at "`000`" as that's what we had in our two `have_a_first_contact` and `have_a_second_contact` steps). The `expected_list` variable is meant to contain the list of contacts, one by line with their phone number. For the "`Mario,Luigi`" example, the content would thus be as follows:

```
Mario 000
Luigi 000
```

Once we have the `expected_list` variable containing the properly formatted text, we only have to compare it to the actual output of the application to confirm that the application printed the two contacts we expected with their phone numbers.

Now that we have translated our steps to code, we can run our test suite to confirm that the test is actually verifying our implementation:

```
$ pytest -v
...
.../test_acceptance.py::test_listing_added_contacts[Mario-Luigi-
Mario,Luigi] FAILED
.../test_acceptance.py::test_listing_added_contacts[John-Jane-John,Jane]
FAILED
...
E ValueError: not enough values to unpack (expected 2, got 1)
```

As expected, since we haven't yet implemented any support for listing contacts, the software crashed, but at least we know that `pytest-bdd` was able to identify the code for all the steps, translate them, and run the scenario for both our examples (as we have the same `test_listing_added_contacts` test performed twice, one for `Mario-Luigi-Mario,Luigi` and one for `John-Jane-John,Jane`).

As usual, we can jump to our functional and unit tests to drive the actual implementation, and a possible edit to our `Application` object could be to handle commands that don't have any `args`, and then call a `printlist` function when the command is "`ls`":

```
class Application:
    ...

    def run(self, text):
        text = text.strip()
        _, cmd = text.split(maxsplit=1)
        try:
```

```
        cmd, args = cmd.split(maxsplit=1)
    except ValueError:
        args = None

    if cmd == "add":
        name, num = args.rsplit(maxsplit=1)
        try:
            self.add(name, num)
        except ValueError as err:
            print(err)
            return
    elif cmd == "del":
        self.delete(args)
    elif cmd == "ls":
        self.printlist()
    else:
        raise ValueError(f"Invalid command: {cmd}")
...

def printlist(self):
    for c in self._contacts:
        print(f"{c[0]} {c[1]}")
```

The printlist function simply iterates over all contacts and prints them with their phone numbers.

As we have the implementation in place, our acceptance test should pass and confirm that it behaves like it is meant to:

```
$ pytest -v
...
.../test_acceptance.py::test_listing_added_contacts[Mario-Luigi-
Mario,Luigi] PASSED
.../test_acceptance.py::test_listing_added_contacts[John-Jane-John,Jane]
PASSED
...
```

Now that the acceptance tests pass for the examples we provided, we know that the implementation satisfies what our team wanted so far.

Summary

In this chapter, we saw how we can write acceptance tests that can be shared with other stakeholders to review the behavior of the software and not just be used by developers as a way to verify that behavior. We saw that it's possible to express the specifications of the software itself in the form of scenarios and examples, which guarantees that our specifications are always in sync with what the software actually does and that our software must always match the specifications as they become the tests themselves.

Now that we know how to move a project forward in a test-driven way using PyTest, in the next chapter we are going to see more essential PyTest plugins that can help us during our daily development practice.

8
PyTest Essential Plugins

In the previous chapter, we saw how to work with PyTest and pytest-bdd to create acceptance tests and verify the requirements of our software.

However, pytest-bdd is not the only useful plugin that PyTest has. In this chapter, we are going to continue working on the contacts project introduced in Chapter 7, *Fitness Function with a Contact Book Application*, showing how some of the most commonly used PyTest plugins can help during the development of a project.

The plugins we are going to cover in this chapter are going to help us with verifying our test suite coverage of the application code, checking the performance of our application, dealing with tests that are flaky or unstable, and optimizing our development process by running only the impacted tests when we change the code base or by speeding up our whole test suite execution.

In this chapter, we will cover the following topics:

- Using pytest-cov for coverage reporting
- Using pytest-benchmark for benchmarking
- Using flaky to rerun unstable tests
- Using pytest-testmon to rerun tests on code changes
- Running tests in parallel with pytest-xdist

Technical requirements

We need a working Python interpreter with the PyTest framework installed with the pytest-bdd plugin. PyTest and pytest-bdd can be installed with the following command:

```
$ pip install pytest pytest-bdd
```

For each section, you will need to install the plugin discussed in the section itself. You can install all of them at once:

```
$ pip install pytest-cov pytest-benchmark flaky pytest-testmon pytest-xdist
```

The examples have been written on Python 3.7 with PyTest 6.0.2 and pytest-bdd 4.0.1, but should work on most modern Python versions. The versions of the plugins in use for each section instead are pytest-cov 2.10, pytest-benchmark 2.3.2, flaky 3.7.0, pytest-testmon 1.0.3, and pytest-xdist 2.1.0.

You can find the code files present in this chapter on GitHub at `https://github.com/PacktPublishing/Crafting-Test-Driven-Software-with-Python/tree/main/Chapter08`.

Using pytest-cov for coverage reporting

We have already seen in `Chapter 1`, *Getting Started with Software Testing*, how code coverage by tests is a good measure for establishing how confident you can be in your test suite. A test suite that only runs 10% of all our code is probably not going to be very reliable in finding problems, as most of the code will go unchecked. A test suite that instead verifies 100% of our code is certainly going to exercise every single line of code we wrote and so should trigger bugs more easily if there are any.

Obviously, coverage cannot verify code that you never wrote, so it's not going to detect that you have a bug because you forgot to add an `if` check in your method, but at least it tells you if you forgot to write a test for that method.

Normally, test coverage in Python is done using the `coverage` module, which can be installed from PyPI, but PyTest has a convenient `pytest-cov` plugin that is going to do that for us and make our life simpler when we want to check the coverage of our tests. Like any other Python distribution, we can install `pytest-cov` through `pip`:

```
$ pip install pytest-cov
```

Installing `pytest-cov` makes the coverage reporting available through the `--cov` option. Running PyTest with that option will immediately output the coverage at the end of the test suite and will save it in a `.coverage` file to make it available for later consultation.

By default, running just `pytest --cov` will provide the coverage of every single module that was imported during the execution of your tests (including all libraries and frameworks you used in your application), which is not very helpful. As we only care about the coverage of our own software, it's possible to tell `pytest-cov` which package to report coverage for simply by adding it as an argument to the `--cov` option.

As we care about how much of our contacts application is actually verified by our tests, we are going to run `pytest --cov=contacts` so that we get back coverage information only for the contacts package, which is the one we care about:

```
$ pytest --cov=contacts
================== test session starts ==================
plugins: cov-2.10.1, bdd-4.0.1
collected 23 items

tests/acceptance/test_adding.py .. [ 8%]
tests/functional/test_basic.py ... [ 21%]
tests/unit/test_adding.py ...... [ 47%]
tests/unit/test_application.py ....... [ 78%]
tests/unit/test_persistence.py .. [ 86%]
tests/acceptance/test_delete_contact.py . [ 91%]
tests/acceptance/test_list_contacts.py .. [100%]

----------- coverage: platform linux, python 3.8.2-final-0 -----------
Name                       Stmts  Miss  Cover
--------------------------------------------------
src/contacts/__init__.py   48     1     98%
src/contacts/__main__.py   2      2     0%
--------------------------------------------------

TOTAL 50 3 94%
```

Great! Our tests cover nearly all our code. The `contacts/__init__.py` module, which is the one where we have all the code that implements our contact book app, is covered at `98%`. Out of the 48 lines of code that compose it, there is only one line that isn't covered.

But how can we know which one it is? pytest-cov obviously knows; we just have to tell it to print it out. That's what the `--cov-report` option is made for. If we run `pytest` with the `--cov-report=term-missing` option, it's going to tell us the lines of code that were not covered by tests in each Python file:

```
$ pytest --cov=contacts --cov-report=term-missing
...

----------- coverage: platform linux, python 3.8.2-final-0 -----------
Name                        Stmts Miss Cover Missing
--------------------------------------------------------
src/contacts/__init__.py    48    1    98%   68
src/contacts/__main__.py    2     2    0%    1-3
--------------------------------------------------------
TOTAL                       50    3    94%
```

Here, for example, we know that lines 1 to 3 in `contacts/__main__.py` are not tested. And that's OK, as those just import and invoke `contacts.main()` for the convenience of being able to run our contacts program with `python -m contacts` once installed (`module.__main__` is what Python invokes when you pass a module to the -m option):

```
from . import main

main()
```

We can easily tell pytest-cov to ignore that code by simply adding a `pragma: no cover` comment near to the lines or code block we want to exclude from coverage:

```
from . import main # pragma: no cover

main() # pragma: no cover
```

Now, if we rerun our test suite, we will no longer get complaints about the __main__.py module:

```
$ pytest --cov=contacts --cov-report=term-missing
...

----------- coverage: platform linux, python 3.8.2-final-0 -----------
Name                        Stmts Miss Cover Missing
--------------------------------------------------------
src/contacts/__init__.py    48    1    98%   68
src/contacts/__main__.py    0     0    100%
--------------------------------------------------------
TOTAL                       48    1    98%
```

Only the code in `contacts/__init__.py` still reports uncovered code. This is the module that contains the real code of our application, so the uncovered line probably has to be tested for real. Once we check what that line refers to, we discover that we have not yet tested the `main` function:

```
67   def main():
68       raise NotImplementedError()
```

As we haven't tested it, we never noticed that it still has to be implemented. This means that currently, running our `contacts` module will simply crash:

```
$ python -m contacts
Traceback (most recent call last):
 ...
  File "src/contacts/__init__.py", line 68, in main
    raise NotImplementedError()
NotImplementedError
```

Thanks to coverage pointing, we found out that the `main` function didn't have a test for it. We notice that a piece of our application is still lacking and we can now move to provide a test for it and implement it.

We are going to create a new module in `tests/functional/test_main.py` where we are going to write our test for the `main` function. Our test is going to provide some fake data pre-loaded (we are not really interested in involving I/O here, so let's replace it with a fake implementation) and verify that when the user runs the "`contacts ls`" command from the command line, the contacts are actually listed back:

```
import sys
from unittest import mock

import contacts

class TestMain:
    def test_main(self, capsys):
        def _stub_load(self):
            self._contacts = [("name", "number")]

        with mock.patch.object(contacts.Application, "load",
                               new=_stub_load):
            with mock.patch.object(sys, "argv", new=["contacts", "ls"]):
                contacts.main()

        out, _ = capsys.readouterr()
        assert out == "name number\n"
```

The implementation required to pass our test is actually pretty short. We just have to create the application, load the stored contacts, and then run the command provided on the command line:

```
def main():
    import sys

    a = Application()
    a.load()
    a.run(' '.join(sys.argv))
```

We can then verify that we finally have 100% coverage of our code from tests and that they all pass by rerunning the `pytest --cov=contacts` command:

```
$ pytest --cov=contacts
collected 24 items

tests/acceptance/test_adding.py .. [ 8%]
tests/functional/test_basic.py ... [ 20%]
tests/functional/test_main.py . [ 25%]
tests/unit/test_adding.py ...... [ 50%]
tests/unit/test_application.py ....... [ 79%]
tests/unit/test_persistence.py .. [ 87%]
tests/acceptance/test_delete_contact.py . [ 91%]
tests/acceptance/test_list_contacts.py .. [100%]

----------- coverage: platform linux, python 3.8.2-final-0 -----------
Name                         Stmts  Miss  Cover
-----------------------------------------------
src/contacts/__init__.py     51     0     100%
src/contacts/__main__.py     0      0     100%
-----------------------------------------------
TOTAL                        51     0     100%
```

If we want our coverage to be verified on every test run, we could leverage the `addopts` option in `pytest.ini` and make sure that coverage is performed every time we run PyTest:

```
[pytest]
addopts = --cov=contacts --cov-report=term-missing
```

As we have already seen, using `addopts` ensures that some options are always provided on every PyTest execution. Thus, we will add coverage options every time we run PyTest.

Coverage as a service

Now that all our tests are passing and our code is fully verified, how can we make sure we don't forget about verifying our coverage when we extend our code base? As we have seen in `Chapter 4`, *Scaling the Test Suite*, there are services that enable us to run our test suite on every new commit we do. Can we leverage them to also make sure that our coverage didn't worsen?

Strictly speaking, ensuring that the coverage doesn't decrease requires comparing the current coverage with the one of the previous successful run, which is something that services such as Travis CI are not able to do as they don't persist any information after our tests have run. So, the information pertaining to the previous runs is all lost.

Luckily, there are services such as Coveralls that integrate very well with Travis CI and allow us to easily get our coverage in the cloud:

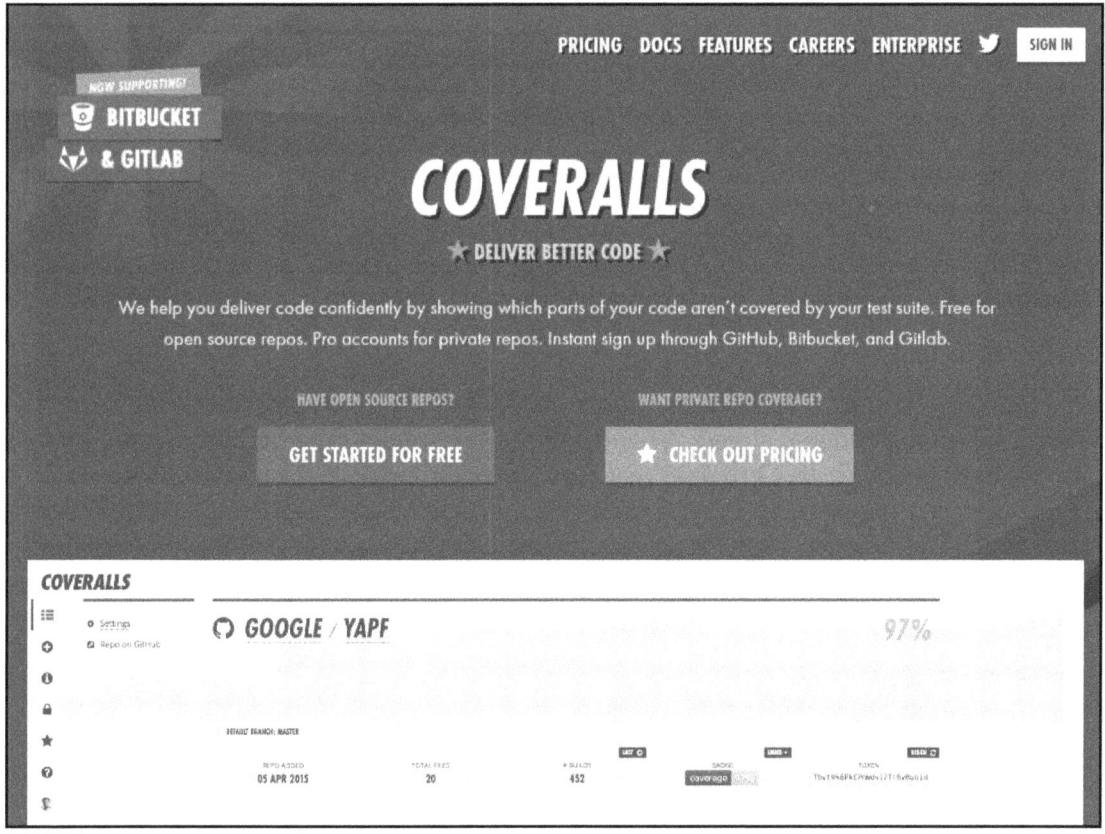

Figure 8.1 – Coveralls web page

As for Travis CI, we can log in with our GitHub account and add any repository that we had on GitHub:

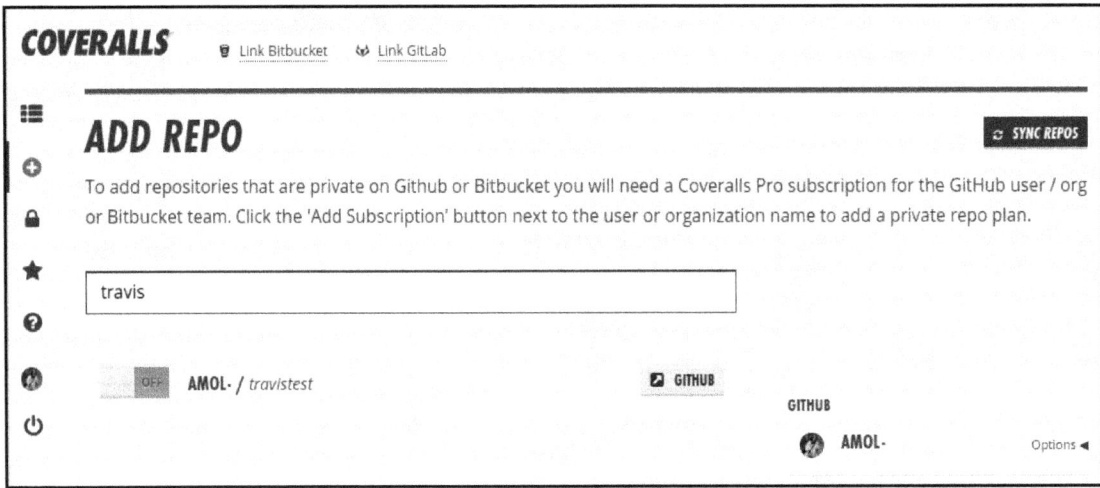

Figure 8.2 – Adding a repo on Coveralls

Once a repository is enabled, Coveralls is ready to receive coverage data for that repository. But how can we get the coverage there?

First of all, we have to tell Travis CI to install support for Coveralls, so, in the install section of our project, `.travis.yml`, we can add the relevant command:

```
install:
  - "pip install coveralls"
```

Then, given that we should already be generating the coverage data by running `pytest --cov`, we have to tell Travis CI to send that data to Coveralls when the test run succeeds:

```
after_success:
  - coveralls
```

Our final `.travis.yml` file should look like the following:

```
install:
  - "pip install coveralls"
  - "pip install -e src"

script:
  - "pytest -v --cov=contacts"
```

```
after_success:
  - coveralls
```

If we have done everything correctly, we should see in Coveralls the trend of our coverage reporting and we should be able to get notified when it lowers or goes below a certain threshold:

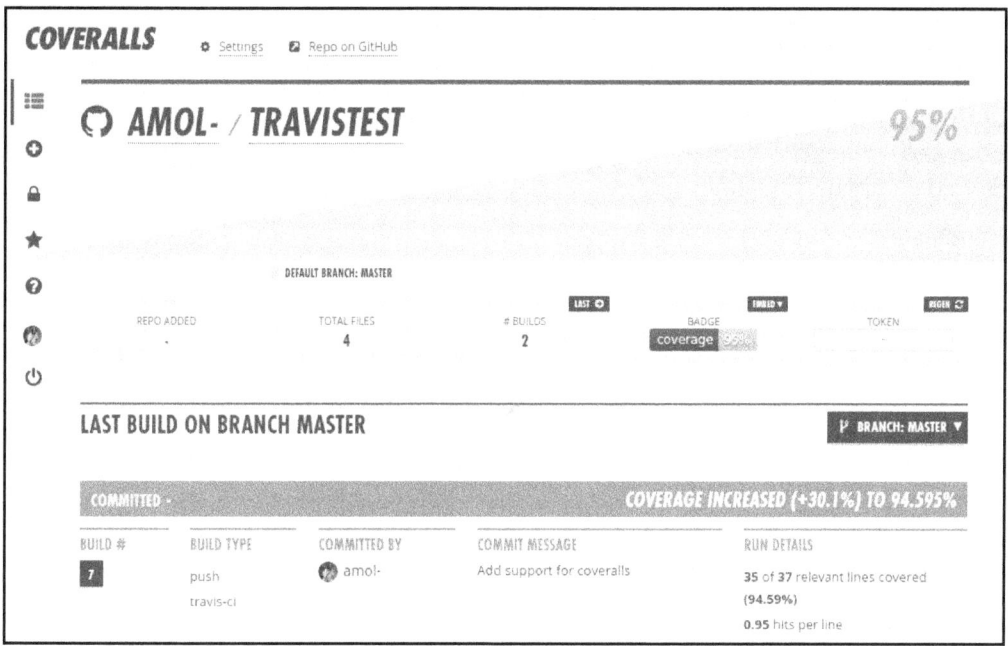

Figure 8.3 – Coveralls coverage reporting

Now that we have our coverage reporting in place, we can move on to taking a look at the other principal plugins that are available for PyTest.

Using pytest-benchmark for benchmarking

Another frequent need when writing applications used by many users is to make sure that they perform in a reasonable way and, hence, that our users don't have to wait too long for something to happen. This is usually achieved by benchmarking core paths of our code base to make sure that slowdowns aren't introduced in those functions and methods. Once we have a good benchmark suite, all we have to do is rerun it on every code change and compare the results to previous runs. If nothing got slower, we are good to go.

PyTest has a `pytest-benchmark` plugin that makes it easy to create and run benchmarks as parts of our test suite. Like any other Python distribution, we can install `pytest-benchmark` through `pip`:

```
$ pip install pytest-benchmark
```

Once we have it installed, we can start organizing our benchmarks in their own dedicated directory. This way, they don't mix with tests, as usually we don't want to run benchmarks on every test run.

For example, if we want to test how fast our app can load 1,000 contacts, we could create a `benchmarks/test_persistence.py` module as the home of a `test_loading` function meant to benchmark the loading of contacts:

```python
from contacts import Application


def test_loading(benchmark):
    app = Application()
    app._contacts = [(f"Name {n}", "number") for n in range(1000)]
    app.save()

    benchmark(app.load)
```

The `benchmark` fixture is provided automatically by `pytest-benchmark` and should be used to invoke the function we want to benchmark, in this case, the `Application.load` method. What your test does is create a new contacts application, and then populates it with a list of 1,000 contacts and saves those contacts on this list. This ensures that we have local contacts to load back.

Then, we can benchmark how long it takes to load back those same contacts, as `benchmark(app.load)` is going to invoke `app.load()`, measuring how long it takes to run it. To run our benchmarks, we can just run them like any other PyTest suite. Running `pytest benchmarks` is enough to get our benchmarks report:

```
$ pytest -v benchmarks
...
benchmark: 3.2.3 (defaults: timer=time.perf_counter disable_gc=False
min_rounds=1 min_time=0.000005 max_time=1.0 calibration_precision=10
warmup=False warmup_iterations=100000)

benchmarks/test_persistence.py::test_loading PASSED [100%]


----------------------- benchmark: 1 tests ----------------------
Name (time in us)    Min      Max         Mean   ...    OPS (Kops/s)    Rounds
-----------------------------------------------------------------
```

```
test_loading          714.7   22,312.3   950.7 ...   1.0518          877
    ------------------------------------------------------------------------

    ==================== 1 passed in 1.96s ====================
```

Running our benchmarks allows us to know that loading back 1,000 contacts takes a minimum of 0.7 milliseconds, a maximum of 22 milliseconds, and an average of 0.9 milliseconds. In total, we can load back 1,000 contacts exactly 1,051 times in a second. pytest-benchmark actually provides much more information about our benchmark run, but for the sake of readability, some of those metrics were excluded in the previously reported run.

How did pytest-benchmark know those metrics? Well, it runs our function 877 times. When dealing with benchmarks, running them only once is usually not enough to get a solid result. If the function is very fast, operative system context switches might weigh on the execution time significantly, and so might provide false results where the time we get is actually heavily influenced by the fact that our system was busy.

pytest-benchmark will decide automatically whether it's necessary to run a benchmark more than once because it's too fast. This is to guarantee that we can get a fairly stable benchmark report even when a very fast function is under the benchmark (and so its execution time can be heavily influenced by system load).

At a minimum, pytest-benchmark will run a function five times before declaring how fast it is. If we have very slow benchmarks and we want them to run no more than once, we can provide the --benchmark-min-rounds=1 option.

Comparing benchmark runs

Now that we know how to run benchmarks, we need to be able to understand whether they got slower compared with previous runs. This can be done by providing --benchmark-autosave --benchmark-compare options to PyTest.

The --benchmark-autosave option will make sure that every benchmark run we perform gets saved in a .benchmarks directory. This way, they are all available for future reference, and then the --benchmark-compare option will tell pytest-benchmark to compare the current run to the one saved previously.

This is a convenient built-in functionality compared to coverage reporting where, in order to ensure non-decreasing coverage, we had to rely on an additional service or implement the check ourselves.

The result of running with `--benchmark-compare` is a report where both runs are provided for comparison:

```
------------------------- benchmark: 2 tests -------------------------
Name (time in us)          Min           Max           Mean         ...
----------------------------------------------------------------------
test_loading (0002_371810a) 726.9 (1.0)  23,884 (1.0)  956.7 (1.0)  ...
test_loading (NOW)          730.1 (1.00) 24,117 (1.01) 969.6 (1.01) ...
----------------------------------------------------------------------
```

For example, in this example, we can see that the previous run (`0002_371810a`) is as fast as the current one (`NOW`), so our code didn't get any slower. If our code base did get slower, pytest-benchmark doesn't only tell us that the performance worsened. It also allows us to know what the bottleneck in our code base is by using the `--benchmark-cprofile=tottime` option.

For example, running our loading benchmark with `--benchmark-cprofile=tottime` will tell us that, as expected, the majority of the time in our `Application.load` function is actually spent reading JSON:

```
test_persistence.py::test_loading (NOW)
ncalls tottime percall cumtime percall filename:lineno(function)
1       0.0004  0.0004  0.0004  0.0004  .../json/decoder.py:343(raw_decode)
1       0.0002  0.0002  0.0002  0.0002  contacts/__init__.py:43(<listcomp>)
1       0.0001  0.0001  0.0009  0.0009  contacts/__init__.py:40(load)
```

Thanks to the performance tests, we have a good understanding of how quick our application can load contacts and where the time loading contacts is spent. This should allow us to evolve it while making sure we don't stray too far from the current performance.

Using flaky to rerun unstable tests

A problem that developers frequently start encountering with fairly big projects that need to involve third-party services, networking, and concurrency is that it becomes hard to ensure that tests that integrate many components behave in a predictable way.

Sometimes, tests might fail just because a component responded later than usual or a thread moved forward before another one. Those are things our tests should be designed to prevent and avoid by making sure the test execution is fully predictable, but sometimes it's not easy to notice that we are testing something that exhibits unstable behavior.

For example, you might be writing an end-to-end test where you are loading a web page to click a button, but at the time you try to click the button, the button itself might not have appeared yet.

Those kinds of tests that sometimes fail randomly are called "flaky" and are usually caused by a piece of the system that is not under the control of the test itself. When possible, it's usually best to put that part of the system under control of the test or replace it with a fake implementation that can be controlled. But when it's not possible, the best we can do is to retry the test.

The `flaky` plugin does that for us. It will automatically retry tests that fail until they pass or up to a maximum number of attempts. An example of such tests is when concurrency is involved. For example, we might write a function that appends entries to a list using threading:

```
def flaky_appender(l, numbers):
    from multiprocessing.pool import ThreadPool

    with ThreadPool(5) as pool:
        pool.map(lambda n: l.append(n), numbers)
```

The test for such a function would probably just check that all the items provided are correctly appended to the list:

```
def test_appender():
    l = []
    flaky_appender(l, range(7000))
    assert l == list(range(7000))
```

Running the test would probably succeed most of the time:

```
$ pytest tests/unit/test_flaky.py -q
tests/unit/test_flaky.py::test_appender PASSED
```

So we might think that our function works OK, but then we start seeing that sometimes, the test fails for no apparent reason.

At this point, we can install the `flaky` plugin to handle our flaky test:

```
$ pip install flaky
```

The first thing we can do is to confirm whether our test is actually flaky by running it multiple times in a row and checking whether it always succeeds. That's something the flaky plugin can do for us through the --min-passes option:

```
$ pytest test_flaky.py --force-flaky --min-passes=10 --max-runs=10

test_appender failed; it passed 9 out of the required 10 times.
        <class 'AssertionError'>
        assert [0, 1, 2, 3, 4, 5, ...] == [0, 1, 2, 3, 4, 5, ...]
   At index 5345 diff: 5600 != 5345
```

As expected, our test succeeded on nine runs, but then failed on the 10th, which confirms that it's a flaky test.

Every time it fails, our entire release process is blocked and we have to rerun the tests and wait for them to complete again. If this happens frequently, it can get frustrating. That's where flaky becomes handy. We can decorate the test with the @flaky decorator to mark it as a flaky test:

```
from flaky import flaky

@flaky
def test_appender():
    l = []
    flaky_appender(l, range(7000))
    assert l == list(range(7000))
```

Now that our test is marked as a flaky one, whenever pytest fails to run it, it will simply retry it, twice by default, but we can control it with the --max-runs option:

```
$ pytest tests/unit/test_flaky.py -v
test_appender failed (1 runs remaining out of 2).
...
test_appender passed 1 out of the required 1 times. Success!
```

In the previous code snippet, our test failed the first run, but flaky noticed that it still had one more try to go out of the default figure of two and retried. Then, on the second try, the test succeeded and PyTest continued.

This allows us to quarantine our flaky tests. We can mark them as flaky and have them not block our release process while we work on providing a more complete solution.

It's usually a good idea to immediately mark as flaky any test that we see unexpectedly fail even just once (unless it's due to a real bug) and then have some dedicated time at which we go through all our flaky tests, trying to unflake them by making the tests more predictable.

Some people prefer to skip the tests that they quarantine, but (while being more robust than marking them as `flaky`) this means that you are willing to live with the risk of introducing any bugs those tests were meant to catch. So, flaky is usually a safer solution and the important part is to have some dedicated time to go back to those quarantined tests to fix them.

Using pytest-testmon to rerun tests on code changes

In a fairly big project, rerunning the whole test suite can take a while, so it's not always feasible to rerun all tests on every code change. We might settle for rerunning all tests only when we commit a stable point of the code and run just a subset of them on every code change before we decide whether to commit our changes.

This approach is usually naturally moved forward by developers who tend to pick a single test, a test case, or a subset of tests that can act as canaries for their code changes.

For example, if I'm modifying the persistence layer of our contacts application, I would probably rerun all tests that involve the `save` or `load` keywords:

```
$ pytest -k save -k load --ignore benchmarks -v
...
tests/functional/test_basic.py::TestStorage::test_reload PASSED [ 50%]
tests/unit/test_persistence.py::TestLoading::test_load PASSED [100%]
```

Once those canary tests pass, I would rerun the whole test suite to confirm that I actually haven't broken anything and I can commit the relevant code. If there are issues, I would obviously catch them when I run the full test suite, but on a fairly big project that can take tens of minutes, it's not a convenient way to catch errors, and the earlier I'm able to catch any errors, the faster I'll be at releasing my code as I don't have to wait for the full test suite to run on every change.

In our case, would just rerunning the tests that have the `load` and `save` keyword in them be enough to catch all possible issues and thus require us to rerun the whole test suite only once as we are very confident that it will pass?

Probably not. There are quite a few more tests that invoke the persistence layer and don't have those keywords in their name. Also, we might not always be so lucky as to have a set of keywords we can use to pick a set of canary tests for every change we do. That's where `pytest-testmon` comes in handy.

`pytest-testmon` will build a graph of relationships between all our code functions and then, on subsequent runs, we can tell `testmon` to only run the tests that are influenced by the code we change.

Ensure `testmon` is installed, as follows:

```
$ pip install pytest-testmon
```

We can do the first run of our test suite to build the relationship graph between the code and tests:

```
$ pytest --testmon --ignore=benchmarks
================== test session starts ===================
...
testmon: new DB, environment: default
...
collected 25 items
...
================== 25 passed in 2.67s ===================
```

Then, we can change any function of our persistence layer (for example, let's just add `return None` at the end of the `Application.save` function), as follows:

```
def save(self):
    with open("./contacts.json", "w+") as f:
        json.dump({"_contacts": self._contacts}, f)
    return None
```

And then we can rerun all the tests that are somehow related to saving data by rerunning `testmon` again:

```
$ pytest --testmon --ignore=benchmarks
================== test session starts ===================
...
testmon: new DB, environment: default
...
collected 16 items / 14 deselected / 2 selected

tests/unit/test_persistence.py . [  9%]
tests/unit/test_adding.py ... [ 36%]
tests/acceptance/test_adding.py .. [ 54%]
tests/functional/test_basic.py .. [ 72%]
tests/acceptance/test_list_contacts.py . [ 81%]
tests/acceptance/test_delete_contact.py . [ 90%]
tests/acceptance/test_list_contacts.py . [100%]

=========== 11 passed, 14 deselected in 1.30s ============
```

In this second run, you can see that instead of running all 25 tests that we had, `testmon` only picked 11 of them, those that somehow invoked the `Application.save` method directly or indirectly, in other words, those that might end up being broken by a change to the method.

Every time we rerun `pytest` with the `--testmon` option, only the tests related to the code that we have changed will be rerun. If we try to run `pytest --testmon` again, for example, no tests would be run as we haven't changed anything from the previous run:

```
$ pytest --testmon --ignore=benchmarks
================== test session starts ===================
...
testmon: new DB, environment: default
...
collected 0 items / 25 deselected

================= 25 deselected in 0.14s ==================
```

This is a convenient way to pick only those tests that are related to our recent code changes and to verify our code on every code change without having to rerun the entire test suite or guess which tests might need to be checked again.

It should be remembered, by the way, that if the behavior of the code depends on configuration files or data saved on disk or on a database, then `testmon` can't detect that tests have to be rerun to verify the behavior again when those change. In general, by the way, having your test suite depend on the state of external components is not a robust approach, so it's better to make sure that your fixtures take care of setting up a fresh state on every run.

Running tests in parallel with pytest-xdist

As your test suite gets bigger and bigger, it might start taking too long to run. Even if strategies to reduce the number of times you need to run the whole test suite are in place, there will be a time where you want all your tests to run and act as the gatekeeper of your releases.

Hence, a slow test suite can actually impair the speed at which we are able to develop and release software.

While great care must always be taken to ensure that our tests are written in the fastest possible way (avoid throwing `time.sleep` calls everywhere, they can be very good at hiding themselves in the most unexpected places), slow components of the software that we are testing should be replaced with fake implementations every time so that it is possible that we can get to a point where there isn't much else we can do and making our test suite faster would be too complex or expensive.

When we get to that point, if we wrote our tests such that they are properly isolated (the state of one test doesn't influence or depend on the state of another test), a possible direction to pursue is to parallelize the execution of our tests.

That's exactly what we can achieve by installing the `pytest-xdist` plugin:

```
$ pip install pytest-xdist
```

Once `xdist` is available, our tests can be run using multiple concurrent workers with the `-n numprocesses` option:

```
$ pytest -n 2
=========== test session starts ==========
. . .
gw0 [26] / gw1 [26]
........................ [100%]
============ 26 passed in 2.71s ==========
```

With `-n 2`, two workers were started for our tests (`gw0` and `gw1`) and tests were equally distributed between the two. Nearly half of the tests should have gone to `gw0` and the other half to `gw1` (PyTest doesn't actually divide the tests equally; it depends on how fast they are to run, but in general, anticipating that tests are equally split is a good approximation).

 Note that as benchmarks can't provide reliable results when run concurrently, pytest-benchmark will disable benchmarking when the `-n` option is provided. The benchmarks will run as normal tests, so you might want to just skip them by explicitly pointing PyTest to the tests directory only, or by using `--ignore benchmarks`.

We can see how tests are distributed simply by running `pytest` in verbose mode with `-v`. In verbose mode, near to every test, we will see which worker was in charge of executing the test:

```
. . .
[gw0] [ 12%] PASSED test_adding.py::TestAddingEntries::test_basic
[gw1] [ 16%] PASSED test_main.py::TestMain::test_main
. . .
```

If you are unsure about how many workers to start, the −n option also accepts the value "auto", which will detect how many processes to start based on how many CPUs the system has.

It is, by the way, important to note that if the test suite is very fast and runs in just a matter of seconds, running it in parallel might actually just make it slower. Distributing the tests across different workers and starting them involves some extra work.

Summary

In this chapter, we saw the most frequently used plugins that exist for PyTest, those plugins that can make your life easier by taking charge of some frequent needs that nearly every test suite will face.

But there isn't any PyTest plugin that is able to manage the test environment itself. We are still forced to set up manually all dependencies that the tests have and ensure that the correct versions of Python are available to run the tests.

It would be great if there was a PyTest plugin able to install everything that we need in order to run our test suite and just "run tests" on a new environment. Well, the good news is that it exists; it's not strictly a PyTest plugin, but it's what Tox, which we are going to introduce in the next chapter, was designed for.

Managing Test Environments with Tox

9

In the previous chapter, we covered the most frequently used PyTest plugins. Through them, we are able to manage our test suite within a Python environment. We can configure how the test suite should work, as well as enable coverage reporting, benchmarking, and many more features that make it convenient to work with our tests. But what we can't do is manage the Python environment itself within which the test suite runs.

Tox was invented precisely for that purpose; managing Python versions and the environment that we need to run our tests. Tox takes care of setting up the libraries and frameworks we need for our test suite to run and will check our tests on all Python versions that are available.

In this chapter, we will cover the following topics:

- Introducing Tox
- Testing multiple Python versions with Tox
- Using Tox with Travis

Technical requirements

We need a working Python interpreter along with Tox. Tox can be installed with the following command:

```
$ pip install tox
```

Even though we are going to use the same test suite and contacts app we wrote in Chapter 8, *PyTest Essential Plugins*, we only need to install Tox 3.20.0. All other dependencies will be managed by Tox for us.

You can find the code files present in this chapter on GitHub at
`https://github.com/PacktPublishing/Crafting-Test-Driven-Software-with-Python/tree/main/Chapter09`.

Introducing Tox

Tox is a virtual environment manager for Python. It takes care of creating the environments and installing our project and all its dependencies on multiple Python versions.

It is a convenient tool that can automate the setup of our project environment and abstract it in a way that we can reuse the same command both locally and in our **Continuous Integration (CI)** pipeline to set up our project and run its tests. It also does that on multiple Python versions at the same time, so that we can check that our project works on all of them.

Testing multiple Python versions can be very convenient when you need to upgrade from one version to the next. Before switching all your systems to the new one, you want to ensure that your code is still able to work on both the old and new versions, so that you can perform a phased rollout.

If we take our contacts application example from Chapter 8, *PyTest Essential Plugins*, the test suite required many dependencies to run. We needed `flaky` to manage flaky tests, `pytest-benchmark` for the benchmarks suite, `pytest-bdd` for the acceptance tests, `pytest-cov` to ensure that the code coverage was verified, and obviously `pytest` itself to run the test suite.

If we had to remember to tell all our colleagues working on the same project to install those packages, it would be easy to forget some of them or end up with incorrect versions installed. We could document our test dependencies, but even better would be to have them managed automatically for us.

So, let's create a `tox.ini` file in our project directory, telling Tox where to find the project to test, which dependencies are necessary to run the test suite, and how to run it:

```
[tox]
setupdir = ./src

[testenv]
deps =
    pytest == 6.0.2
    pytest-bdd == 3.4.0
    flaky == 3.7.0
    pytest-benchmark == 3.2.3
```

```
    pytest-cov == 2.10.1
commands =
    pytest --cov=contacts
```

The `[tox]` section configures Tox itself. In this case, it can ascertain through the `setupdir` = option where to find the project that is under test.

The `[testenv]` section is instead meant to provide directives for each environment in which we want to test our project. In this case, through the `deps` = option, we are listing all things that need to be installed in that environment so that the project can be tested (the project itself is always automatically installed by Tox, so no need to list it here), and by using the `commands` = options, we are telling Tox how to test the project in the environments.

Once this file is in place in the root of our project, we can prepare a fully working environment and test the project by simply invoking the `tox` command:

```
$ tox
GLOB sdist-make: ./src/setup.py
python create: ./.tox/python
python installdeps: pytest == 6.0.2, pytest-bdd == 3.4.0, flaky == 3.7.0,
pytest-benchmark == 3.2.3, pytest-cov == 2.10.1
python inst: ./.tox/.tmp/package/1/contacts-0.0.0.zip
python installed: ...
python run-test: commands[0] | pytest --cov=contacts
===================== test session starts ======================
...
collected 26 items

tests/acceptance/test_delete_contact.py . [ 3%]
tests/acceptance/test_list_contacts.py .. [ 11%]
benchmarks/test_persistence.py . [ 15%]
tests/acceptance/test_adding.py .. [ 23%]
tests/functional/test_basic.py ... [ 34%]
tests/functional/test_main.py . [ 38%]
tests/unit/test_adding.py ...... [ 61%]
tests/unit/test_application.py ....... [ 88%]
tests/unit/test_flaky.py . [ 92%]
tests/unit/test_persistence.py .. [100%]

----------- coverage: platform linux, python 3.7.3-final-0 -----------
Name                      Stmts Miss Cover
--------------------------------------------------------------------
contacts/__init__.py      51    0    100%
contacts/__main__.py      0     0    100%
--------------------------------------------------------------------
TOTAL                     51    0    100%
```

```
----------------------- benchmark: 1 tests -------------------------
Name (time in us) Min Max Mean ... OPS (Kops/s) Rounds
--------------------------------------------------------------------
test_loading 714.7 22,312.3 950.7 ... 1.0518 877
--------------------------------------------------------------------


===================== 26 passed in 2.41s =====================
```

As you can see, Tox created a new Python environment in `./tox /python`, installed our project and all the required dependencies for us, and then started the test suite providing coverage and benchmarks.

The side effect of this approach is that we lost a bit of flexibility in terms of what we can tell PyTest. Tox is going to run all our tests and benchmarks. If we only want to run some of them, there is no way of doing this.

This flexibility can be regained by using the Tox `{posargs}` variable, which will proxy all options we provide in the command line from Tox to our test suite. So we can put `{posargs}` in our `commands` option in `tox.ini` so that any additional option we provide to Tox gets forwarded to our test command:

```
commands =
    pytest --cov=contacts {posargs}
```

Now, if we run Tox with any additional option after `--`, it will be forwarded to PyTest. For example, to exclude benchmarks from our run, we can use `tox -- ./tests` to exclude benchmarks and only run the tests that are related to loading back our contacts. Instead, we can use `tox -- ./tests -k load`:

```
$ tox -- ./tests -k load
...
============= test session starts =============
collected 25 items / 23 deselected / 2 selected

tests/functional/test_basic.py . [ 50%]
tests/unit/test_persistence.py . [100%]
...
====== 2 passed, 23 deselected in 0.35s =======
```

Now that we know how to use Tox to set up the testing environment without losing the flexibility that was afforded to us earlier when we did things manually, we can move forward and see how to actually set up multiple testing environments on different versions of Python.

Testing multiple Python versions with Tox

Tox is based on the concept of **environments**. The goal of Tox is to prepare those environments where it will run the commands provided. Usually, those environments are meant for testing (running tests in different conditions) and the most common kind of environments are those that use different Python versions. But in theory, it is possible to create a different environment for any other purpose. For example, we frequently create an environment where project documentation is built.

To add further environments to Tox, it's sufficient to list them inside the envlist = option. To configure two environments that test our project against both **Python 2.7** and **Python 3.7**, we can set envlist to both py37 and py27:

```
[tox]
setupdir = ./src
envlist = py27, py37
```

If we run tox again, we will see that it will now test our project on two different environments, one made for **Python 2.7** and one for **Python 3.7**:

```
$ tox
GLOB sdist-make: ./src/setup.py

py27 create: ./.tox/py27
py27 installdeps: pytest == 6.0.2, pytest-bdd == 3.4.0, flaky == 3.7.0,
pytest-benchmark == 3.2.3, pytest-cov == 2.10.1
...
py37 create: ./.tox/py37
py37 installdeps: pytest == 6.0.2, pytest-bdd == 3.4.0, flaky == 3.7.0,
pytest-benchmark == 3.2.3, pytest-cov == 2.10.1
```

We obviously need to have working executables of those two Python versions on our system, but as far as they are available and running the python3.7 and python2.7 commands works, Tox will be able to leverage them.

By default, all environments apply the same configuration, the one provided in [testenv], so in our case, Tox tried to install the same exact dependencies and run the same exact commands on both **Python 2.7** and **Python 3.7**.

On Python 2.7, it failed because PyTest no longer supports Python 2.7 on versions after 4.6.11, so if we want to actually test our project on Python 2.7, we need to provide a custom configuration for the environment and make it work against a previous PyTest version:

```
py27 create: ./.tox/py27
py27 installdeps: pytest == 6.0.2, pytest-bdd == 3.4.0, flaky == 3.7.0,
pytest-benchmark == 3.2.3, pytest-cov == 2.10.1
ERROR: Could not find a version that satisfies the requirement
pytest==6.0.2 (from versions: 2.0.0, ..., 4.6.11)
ERROR: No matching distribution found for pytest==6.0.2
```

To fix this issue, we can simply go back and provide a custom configuration for the **Python 2.7** environment where we are going to customize the deps = option, stating explicitly that on that version of Python, we want to use a previous PyTest version:

```
[testenv:py27]
deps =
    pytest == 4.6.11
    pytest-bdd == 3.4.0
    flaky == 3.7.0
    pytest-benchmark == 3.2.3
    pytest-cov == 2.10.1
```

Options can be specialized just by creating a section named [testenv:envname], in this case, [testenv:py27], as we want to override the options for the py27 environment.

Any option that isn't specified is inherited from the generic [testenv] configuration, so as we haven't overridden the command = option, the configuration we provided in [testenv] will be used for testing on **Python 2.7**, too.

By running Tox with this new configuration, we will finally be able to set up the environment, install PyTest, and start our tests:

```
$ tox
GLOB sdist-make: ./09_tox/src/setup.py
py27 create: ./09_tox/.tox/py27
py27 installdeps: pytest == 4.6.11, pytest-bdd == 3.4.0, flaky == 3.7.0,
pytest-benchmark == 3.2.3, pytest-cov == 2.10.1
py27 inst: ./.tox/.tmp/package/1/contacts-0.0.0.zip
py27 installed: contacts @
file://./.tox/.tmp/package/1/contacts-0.0.0.zip,pytest==4.6.11,...
py27 run-test-pre: PYTHONHASHSEED='2140925334'
py27 run-test: commands[0] | pytest --cov=contacts
```

As we could have anticipated, our tests fail on Python 2.7 as our project wasn't written to support such an old Python version:

```
platform linux2 -- Python 2.7.16, pytest-4.6.11, py-1.9.0, pluggy-0.13.1
cachedir: .tox/py27/.pytest_cache
rootdir: .
plugins: bdd-3.4.0, flaky-3.7.0, benchmark-3.2.3, cov-2.10.1
collected 5 items / 7 errors

================ ERRORS ===================
    mod = self.fspath.pyimport(ensuresyspath=importmode)
.tox/py27/lib/python2.7/site-packages/py/_path/local.py:704: in pyimport
    __import__(modname)
E File "./benchmarks/test_persistence.py", line 5
E app._contacts = [(f"Name {n}", "number") for n in range(1000)]
E                          ^
E SyntaxError: invalid syntax
========== 7 error in 1.07 seconds ========
```

For example, we used f-strings, which were not supported on Python 2.7. Porting projects to Python 2.7 is beyond the scope of this book, so we are not going to modify our project to make it work there, but the same concepts that we have seen while using Python 2.7 do apply to any other environment.

For example, if, instead of Python 2.7, we wanted to test our project against **Python 3.8**, we could have just used `py38` instead of `py27` as the name of the environment. In that case, we wouldn't even have to customize the `deps` = option for that environment as **PyTest 6** works fine on **Python 3.8**.

Using environments for more than Python versions

By default, Tox provides a few predefined environments for various Python versions, but we can declare any kind of environment that differs for whatever reason.

Another common way to use this capability is to create various environments that differ for the `commands` = option, and so do totally different things. You will probably frequently see that this used to provide a way to build project documentation. It is not uncommon to see a `docs` environment in Tox configurations that, instead of running tests, builds the project documentation.

In our case, we might want to use this feature to disable benchmarks by default and make them run only when a dedicated environment is used.

To do so, we are going to disable benchmarks by default in our [testenv] configuration:

```
[tox]
setupdir = ./src
envlist = py27, py37

[testenv]
deps =
    pytest == 6.0.2
    pytest-bdd == 3.4.0
    flaky == 3.7.0
    pytest-benchmark == 3.2.3
    pytest-cov == 2.10.1
commands =
    pytest --cov=contacts --benchmark-skip {posargs}

[testenv:py27]
...
```

Then we are going to add one more [testenv:benchmarks] environment that runs only the benchmarks:

```
[testenv:benchmarks]
commands =
    pytest --no-cov ./benchmarks {posargs}
```

This environment will inherit the configuration from our default environment, and thus will use the same exact deps, but will provide a custom command where coverage is disabled and only benchmarks are run.

It is important that we don't list this environment in the envlist option of the [tox] section. Otherwise, the benchmarks would end up being run every time we invoke Tox, which is not what we want.

To explicitly run benchmarks on demand, we can run Tox with the -e benchmarks option, which will run Tox just for that specific environment:

```
$ tox -e benchmarks
GLOB sdist-make: ./src/setup.py
benchmarks create: ./.tox/benchmarks
benchmarks installdeps: pytest == 6.0.2, pytest-benchmark == 3.2.3, ...
benchmarks inst: ./.tox/.tmp/package/1/contacts-0.0.0.zip
benchmarks run-test-pre: PYTHONHASHSEED='257991845'
benchmarks run-test: commands[0] | pytest --no-cov ./benchmarks
======================= test session starts =======================
platform linux -- Python 3.7.3, pytest-6.0.2, py-1.9.0, pluggy-0.13.1
collected 1 item
```

```
benchmarks/test_persistence.py .                        [100%]


------------------------ benchmark: 1 tests ------------------------
Name (time in us) Min Max Mean ... OPS (Kops/s) Rounds
--------------------------------------------------------------------
test_loading 714.7 22,312.3 950.7 ... 1.0518 877
--------------------------------------------------------------------


====================== 1 passed in 1.73s  ========================
```

We now have a configuration where running `tox` by default will run our tests on **Python 2.7** and **Python 3.7**, and then running `tox -e benchmarks` does run benchmarks.

If we further want to specialize the behavior of our Tox configuration, we can do so by adding more environments and customizing the options we care about. A complete reference of all the Tox options is available on the **ReadTheDocs** page of Tox, so make sure to take a look if you want to dive further into customizing Tox behavior.

Now that we have Tox working locally, we need to combine it with our CI system to ensure that different CI processes are started for each Tox environment. As we have used Travis for all our CI needs so far, let's see how we can integrate Tox with Travis.

Using Tox with Travis

Using Tox with a CI environment is usually fairly simple, but as both Tox and the CI will probably end up wanting to manage the Python environment, some attention has to be paid to enable them to exist together. To see how Travis and Tox can work together, we can pick our chat project that we wrote in Chapter 4, *Scaling the Test Suite*, which we already had on Travis-CI, and migrate it to use Tox.

We need to write a `tox.ini` file, which will take care of running the test suite itself:

```
[tox]
setupdir = ./src
envlist = py37, py38, py39

[testenv]
usedevelop = true
deps =
    coverage
commands =
    coverage run --source=src -m unittest discover tests -v
    coverage report
```

```
[testenv:benchmarks]
commands =
    python -m unittest discover benchmarks
```

The commands you see in tox.ini are the same that we previously had in the travis.yml file under the script: section. That's because, previously, Travis itself was in charge of running our tests. Now, Tox will be in charge of doing so.

For the same reason, as the coverage reporting should happen every time we run the test suite, we have Tox install the coverage dependency and run coverage report after the test suite.

The main difference with tox.ini seen previously in the chapter is probably the usedevelop = true option. That tells Tox to install our own project in editable mode (sometimes called developer mode). Instead of making a distribution package out of our source directory and then installing the distribution, Tox will install the source directory itself. This is frequently convenient when coverage reporting is involved as we usually want the coverage to be against our source code, and not against the installed distribution.

The benefit of using a Tox file is that it should work the same everywhere. So, before moving it to Travis, we can verify that it actually does what we expect locally on our own machine:

```
$ tox
py38 develop-inst-noop: travistest/src
py38 run-test: commands[0] | coverage run --source=src -m unittest discover
tests -v
test_message_exchange (e2e.test_chat.TestChatAcceptance) ... ok
test_smoke_sending_message (e2e.test_chat.TestChatAcceptance) ... ok
test_exchange_with_server (functional.test_chat.TestChatMessageExchange)
... ok
test_many_users (functional.test_chat.TestChatMessageExchange) ... ok
test_multiple_readers (functional.test_chat.TestChatMessageExchange) ... ok
test_client_connection (unit.test_client.TestChatClient) ... ok
test_client_fetch_messages (unit.test_client.TestChatClient) ... ok
test_nickname (unit.test_client.TestChatClient) ... ok
test_send_message (unit.test_client.TestChatClient) ... ok
test_broadcast (unit.test_connection.TestConnection) ... ok


----------------------------------------------------------------------
Ran 10 tests in 0.058s

OK
py38 run-test: commands[1] | coverage report
Name Stmts Miss Cover
----------------------------------------
```

```
src/chat/__init__.py    0 0 100%
src/chat/client.py     29 0 100%
src/chat/server.py      7 0 100%
src/setup.py            2 2 0%
----------------------------------------
TOTAL                  38 2 95%
```

As desired, it ran the test suite and then reported the code coverage. We also know, thanks to [testenv:benchmarks], that if we want, we can run benchmarks with tox -e benchmarks:

```
$ tox -e benchmarks
benchmarks develop-inst-noop: travistest/src
benchmarks run-test: commands[0] | python -m unittest discover benchmarks

  time: 0.06, iteration: 0.01

.
-----------------------------------------------------------------
Ran 1 test in 0.069s

OK
```

Now, the remaining element is to make Tox run inside Travis.

To do so, mostly we have to replace the script: section in our travis.yml file with a single tox command. Then, Tox will do everything it has to do in order to make the tests run as it did on our own PC:

```
script:
  - "tox"
```

However, Travis will also need Tox itself to run the commands. Therefore, we want to have Travis install Tox before running the script. To do so, we are going to use a special package named tox-travis and we are going to add it to the install: section:

```
install:
  - "pip install tox-travis"
```

You might be wondering why we used tox-travis instead of just tox. The reason is that tox-travis takes care of that little extra work that is necessary to make Tox and Travis collaborate. By default, Travis wants to install and set up Python, but Tox also wants to do the same. That means that we would end up installing Python twice.

Even worse, as we have `envlist = py37, py38, py39` in our `tox.ini`, Tox would actually try to run the tests against all three Python versions for each Travis Python environment. So, suppose that we asked Travis to set up 3.7, 3.8, and 3.9. Then, Tox would try to use 3.7, 3.8, and 3.9 inside the Travis 3.7 Python environment, and would then try to use 3.7, 3.8, and 3.9 inside the Travis 3.8 Python environment, and so on, leading to errors such as the following:

```
ERROR: py38: InterpreterNotFound: python3.8
ERROR: py39: InterpreterNotFound: python3.9
```

To avoid this problem, we can use `tox-travis`. When we use Tox-Travis, the Python environments come from Travis only and Tox will simply use those already prepared by Travis without trying to set up a second Python environment. At that point, our Tox `envlist` is only helpful locally, and on Travis, the `python:` section of the `travis.yml` file will dictate which Python versions get used.

Apart from making sure that we install `tox-travis`, the rest of our `travis.yml` file is fairly similar to the original one our project had previously. We just replaced the commands to run tests and benchmarks with those that Tox provides:

```
language: python

os: linux
dist: xenial

python:
  - 3.7
  - &mainstream_python 3.8
  - 3.9
  - nightly

install:
  - "pip install tox-travis"
  - "pip install coveralls"

script:
  - "tox"

after_success:
  - coveralls
  - "tox -e benchmarks"
```

Now that both our `tox.ini` and `travis.yml` configuration files are in place, we can just push our repository changes and see that Travis successfully runs our tests using Tox:

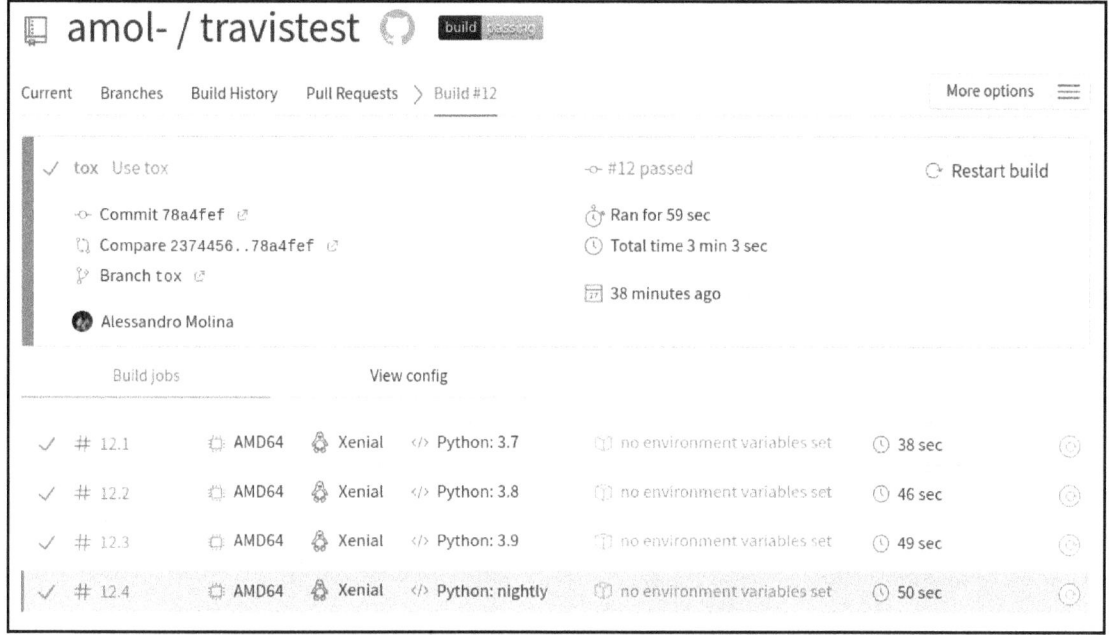

Figure 9.1 – Tox setup

It should became clear that once we have a working local Tox setup, moving on to Travis involves very little apart from writing a `travis.yml` configuration file in charge of installing `tox-travis` and then invoking `tox`.

Summary

In this chapter, we saw how Tox can take care of all the setup necessary to run our tests for us and how it can do that on multiple target environments so that all we have to do to run tests is just to invoke Tox itself.

This is a more convenient, but also robust, way to manage our test suite. The primary benefit is that anyone else willing to contribute to our project won't have to learn how to set up our projects and how to run tests. If our colleagues or project contributors are familiar with Tox, seeing that our project includes a `tox.ini` file tells them all that they will need to know—that they just have to invoke the `tox` command to run tests.

Now that we have seen the base plugins and tools to manage and run our test suite, in the next chapter, we can move on to some more advanced topics that involve how to test our documentation itself and how to use property-based testing to catch bugs in our code.

10
Testing Documentation and Property-Based Testing

In the previous chapter, we saw how to manage the environment where the test suite runs through Tox. We now have a fairly good understanding of how to create a test suite, how to set up an environment where this can be run, and how to ensure that we are able to organize it in a way that remains effective as our software and test suite grow. We are now going to move our attention to confirm that our tests are able to identify and cover corner cases and make sure that our documentation is as robust and tested as our software itself.

In this chapter, we will cover the following topics:

- Testing documentation
- Property based-testing

Technical requirements

We need a working Python interpreter with PyTest, Sphinx for documentation testing, and the Hypothesis framework for property-based testing. All of them can be installed through pip with the help of the following command:

```
$ pip install pytest sphinx hypothesis
```

The examples have been written on Python 3.7, Sphinx 3.3.0, PyTest 6.0.2, and Hypothesis 5.41, but should work on most modern Python versions.

You can find the code files present in this chapter on GitHub at `https://github.com/PacktPublishing/Crafting-Test-Driven-Software-with-Python/tree/main/Chapter10`.

Testing documentation

When documentation is written with the goal of teaching other developers how a system works, providing examples on how to use its inner layers, and train them on the driving design principles behind some complex software, it can be a very effective way to onboard new team members in a project.

In any fairly big and complex project, documentation becomes something that is essential for navigating the complexity of the system without having to rely on our memory to remember how to use every single layer or class involved in the system.

But documentation is also hard. Not only is it actually hard to write, because what might seem obvious and clear to us might sound cryptic to another reader, but also because the code evolves quickly and documentation easily becomes outdated and inaccurate.

Thankfully, testing is a very effective way to also ensure that our documentation doesn't get outdated and that it still applies to our system. As much as we test the application code, we can test the documentation examples. If an example becomes outdated, it will fail and our documentation tests won't pass.

Given that we have covered every human-readable explanation in our documentation with a code example, we can make sure that our documentation doesn't get stale and still describes the current state of the system by verifying those code examples. To show how documentation can be kept in sync with the code, we are going to take our existing contacts application we built in previous chapters and we are going to add tested documentation to it.

Our first step will be to create the documentation itself. In Python, the most common tool for documentation is `Sphinx`, which is based on the **reStructuredText** format.

Sphinx provides the `sphinx-quickstart` command to create new documentation for a project. Running `sphinx-quickstart docs` will ask a few questions about the layout of our documentation project and will create it inside the `docs` directory. We will also provide the `--ext-doctest --ext-autodoc` options to enable the extensions to make documentation testable and to autogenerate documentation from existing code:

```
$ sphinx-quickstart docs --ext-doctest --ext-autodoc
Welcome to the Sphinx 3.3.0 quickstart utility.

...

> Separate source and build directories (y/n) [n]: y
> Project name: Contacts
> Author name(s): Alessandro Molina
```

```
> Project release []:
> Project language [en]:

Creating file docs/source/conf.py.
Creating file docs/source/index.rst.
Creating file docs/Makefile.
Creating file docs/make.bat.

Finished: An initial directory structure has been created.
```

Once our documentation is available in the `docs` directory, we can start populating it, beginning with `docs/source/index.rst`, which will be the entry point for our documentation. If we want to add further sections to it, we have to list them under the `toctree` section.

In our case, we are going to create a section about how to use the software and a reference section for the existing classes and methods. Therefore, we are going to add `contacts` and `reference` sections to `toctree` in the `docs/source/index.rst` file:

```
Welcome to Contacts's documentation!
====================================

.. toctree::
   :maxdepth: 2
   :caption: Contents:

   contacts
   reference
```

Now, we could try to build our documentation to see whether the two new sections are listed on the home page. But doing so would actually fail because we haven't yet created the files for those two sections:

```
$ make html
Running Sphinx v3.3.0
...
docs/source/index.rst:9: WARNING: toctree contains reference to nonexisting
document 'contacts'
docs/source/index.rst:9: WARNING: toctree contains reference to nonexisting
document 'reference'
```

So, we are going to create `docs/source/contacts.rst` and `docs/source/reference.rst` files to allow Sphinx to find them.

Adding a code-based reference

First, we will add the reference section, as it's the simplest one. The `docs/source/reference.rst` file will only contain the title and the directive that will tell Sphinx to document the `contacts.Application` class based on the docstring we provide in the code itself:

```
==============
Code Reference
==============

.. autoclass:: contacts.Application
   :members:
```

Recompiling our documentation with `make html` will now only report the missing `contacts.rst` file and successfully generate the code reference section. The result will be visible in the `docs/build/` directory, hence, opening the `docs/build/reference.html` file will now show our code reference.

The first time we build it, our reference will be mostly empty:

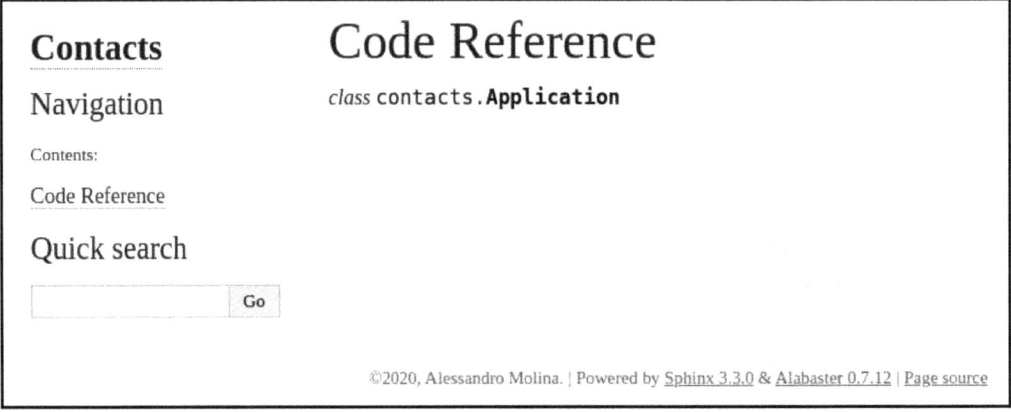

Figure 10.1 – Code reference

It has a section for the `contacts.Application` class, but nothing else. This is because the content is taken directly from the code docstrings, and we haven't written any.

Therefore, we should go back to our `contacts/__init__.py` file and add a docstring to
our `Application` class and to the `Application.run` method:

```
class Application:
    """Manages a contact book serving the provided commands.

    The contact book data is saved in a contacts.json
    file in the directory the application is
    launched from. Any contacts.json in the directory this
    is launched from will be loaded at init time.

    A contact is composed by any name followed by a valid
    phone number.
    """
    PHONE_EXPR = re.compile('^[+]?[0-9]{3,}$')

    def __init__(self):
        self._clear()

    def _clear(self):
        self._contacts = []

    def run(self, text):
        """Run a provided command.

        :param str text: The string containing the command to run.

        Takes the command to run as a string as it would
        come from the shell, parses it and runs it.

        Each command can support zero or multiple arguments
        separate by an empty space.

        Currently supported commands are:

          - add
          - del
          - ls
        """
        ...
```

Now that the class and the method are both documented, we can rebuild our
documentation with `make html` to see whether the reference has been properly generated.

If everything works as expected, we should see in `docs/build/reference.html` the documentation we wrote in the code:

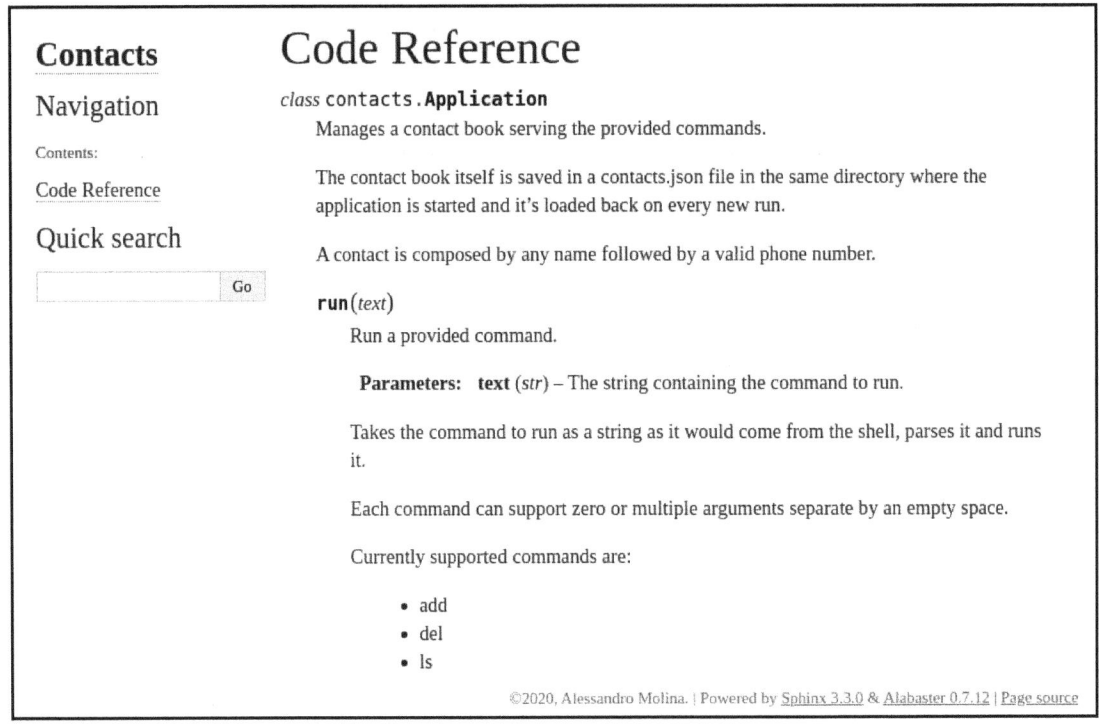

Figure 10.2 – Reference generated

Mixing code and documentation in the source files is an effective technique for ensuring that when the code changes, the documentation is updated too. For example, if we remove a method, we would surely also remove its docstring too, and so the method would also disappear from the documentation. Obviously, we still have to pay attention that what we write in the docstrings makes sense, but at least the structure of our documentation would always be in sync with the structure of our code.

Writing a verified user guide

While it's effective for references, having a reference is usually far from being enough for proper documentation. A usage guide and tutorials are frequently necessary to ensure that the reader understands how the software works.

So, to make our documentation more complete, we are going to add a user guide to the `docs/source/contacts.rst` file.

After a brief introduction, the `docs/source/contacts.rst` file will dive into some real-world examples regarding how to add new contacts and how to list them:

```
===============
Manage Contacts
===============

.. contents::

Contacts can be managed through an instance of
:class:`contacts.Application`, use :meth:`contacts.Application.run`
to execute any command like you would in the shell.

Adding Contancts
================

.. code-block::

    app.run("contacts add Name 0123456789")

Listing Contacts
================

.. code-block::

    app.run("contacts ls")
```

Now, if we rebuild our documentation with `make html`, we should no longer get any error and opening `docs/build/contacts.html` should show the page we just wrote with the two examples:

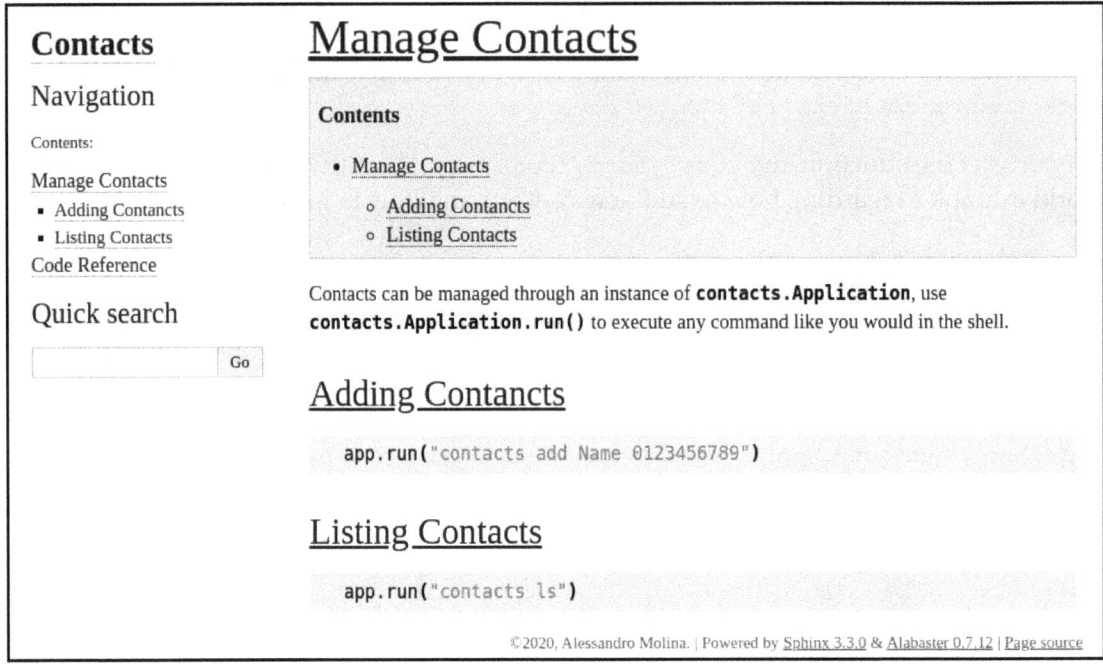

Figure 10.3 – Managing contacts

While this shows how we can use the application, it doesn't do anything to ensure that the documentation is in sync with our code. If, for example, we ever replace the `Application.run` method with `Application.execute`, the two examples on the page won't even notice and will continue to say that you have to use `app.run`, which will be incorrect.

How can we make sure that the examples and tutorials we write are actually always in sync with how our application works for real? That's exactly what we can do using `doctest`. Doctest is a Python module and Sphinx extension that allows us to write snippets of code that are tested and verified in our documentation. So, we are going to use `doctest` to make sure that those two examples actually run and do what we expect.

The first thing we have to do is to set up the application in the documentation file. So we are going to add a `testsetup` directive to `docs/source/contacts.rst` with the code that is necessary to make sure that the `app` object exists for real.

For the sake of order, we are going to add this code at the end of the introductory paragraph, right before the examples themselves:

```
Manage Contacts
===============

.. contents::

Contacts can be managed through an instance of
:class:`contacts.Application`, use :meth:`contacts.Application.run`
to execute any command like you would in the shell.

.. testsetup::

    from contacts import Application
    app = Application()
```

Then we are going to replace the two `code-block` directives with two `testcode` directives, which means that the examples will actually be executed and checked to ensure that they are not crashing:

```
Adding Contacts
===============

.. testcode::

    app.run("contacts add Name 0123456789")

Listing Contacts
===============

.. testcode::

    app.run("contacts ls")
```

`code-block` directives instruct Sphinx that the content should be formatted as code, but does nothing to ensure that the content is actually valid code that does not crash. While the `testcode` directive formats the code, it also ensures that it is valid code that can run.

Now we are verifying that the two commands can actually run, so if we ever renamed `Application.run` to `Application.execute`, our `testcode` examples would fail to run and so Sphinx would complain that we have to update the documentation.

But making sure that they can run is not enough. We also want to ensure that they actually do what we expect, that once we add a contact and list them back, we do see the new contact. The `doctest` module provides us with the `testoutput` directive to ensure that the previous `testcode` block provided the expected output. In this case, we are going to add a `testoutput` directive right after the code block that lists our contacts that will ensure that the contact we just added is listed back:

```
Listing Contacts
================

.. testcode::

    app.run("contacts ls")

.. testoutput::

    Name 0123456789
```

If we rerun `make html`, we are going to see that in the resulting documentation, not much has changed. There is an extra paragraph with the output after the second example, which is good, as it gives a hint of what the expected output of the `ls` command is, but apart from that, our documentation looks the same as before:

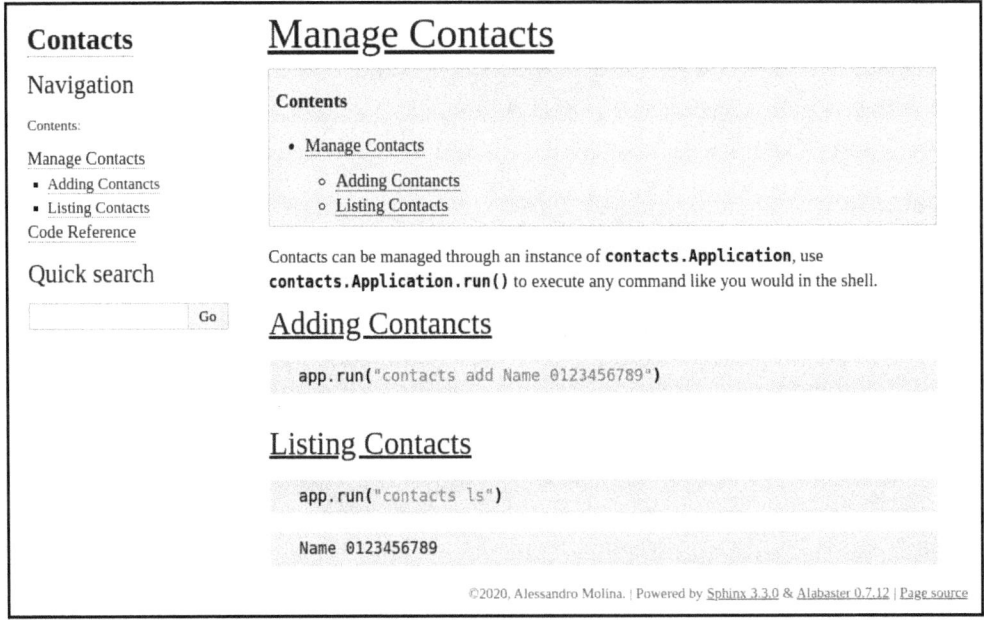

Figure 10.4 – Manage Contacts updated

The real difference happens when we run the `make doctest` command, which allows us to verify that the examples in our documentation do work correctly:

```
$ make doctest
Running Sphinx v3.3.0
...
running tests...

Document: contacts
------------------
1 items passed all tests:
   2 tests in default
2 tests in 1 items.
2 passed and 0 failed.
Test passed.

Doctest summary
===============
    2 tests
    0 failures in tests
    0 failures in setup code
    0 failures in cleanup code
build succeeded.
```

`doctest` found two tests (the two `testcode` blocks) within the `contacts.rst` document and it confirmed that both of them work correctly.

If, as we mentioned before, we ever rename the `Application.run` method to `Application.execute`, the doctests will immediately point out that both examples are wrong:

```
Document: contacts
------------------
**************************************************************************
File "contacts.rst", line 41, in default
Failed example:
    app.run("contacts add Name 0123456789")
Exception raised:
    Traceback (most recent call last):
      File "/usr/lib/python3.8/doctest.py", line 1336, in __run
        exec(compile(example.source, filename, "single",
      File "<doctest default[0]>", line 1, in <module>
        app.run("contacts add Name 0123456789")
    AttributeError: 'Application' object has no attribute 'run'
**************************************************************************
File "contacts.rst", line 55, in default
Failed example:
```

```
        app.run("contacts ls")
Exception raised:
    Traceback (most recent call last):
      File "/usr/lib/python3.8/doctest.py", line 1336, in __run
        exec(compile(example.source, filename, "single",
      File "<doctest default[0]>", line 1, in <module>
        app.run("contacts ls")
    AttributeError: 'Application' object has no attribute 'run'
**************************************************************************
1 items had failures:
   2 of 2 in default
2 tests in 1 items.
0 passed and 2 failed.
***Test Failed*** 2 failures.
```

Likewise, if anything goes wrong in our two examples or the contacts listed don't match those in the `testoutput` section, the `make doctest` command would report those failures and would inform us that our documentation is not in sync with our code.

Adding the `make doctest` command to our CI pipeline allows us to ensure that with every change of code that affects the documentation, the documentation is updated too, thereby guaranteeing that all our examples in the documentation are verified and up to date with what our code actually does.

Property-based testing

Now that we know how to have working test suites for both our code and our documentation, the quality of those test suites fully depends on our capability to design and write good tests.

There is, by the way, one rule in software testing that can help us design good tests, and this is that errors usually hide in corner cases and limit values. If we have a function that performs division between two numbers, the bugs are probably going to be brought to the surface when zero, the maximum integer value, or negative numbers are passed to the function as arguments. Rarely will we see errors for most common values, such as 2, 3, 4, or 5. That's because developers usually tend to design their code with those common values in mind. The design that comes more naturally is usually the one that works for the most obvious cases, and corner cases rarely come to mind in the first instance.

Property-based testing comes in handy when easily generating tests that verify those corner cases and limit conditions by ensuring that some properties of the functions and methods we test always hold true. Property-based testing had its origins in the Haskell community, but libraries and frameworks to implement it are now available in most programming languages, including Python.

Hypothesis is a library that allows us to implement property-based testing in Python.

An example of the properties of a function could be that "for any provided argument, the function should never crash." Not crashing is the most frequently verified property, but it's possible to check any invariant that our method should guarantee. If we have a function such as `concat(a: str, b: str, c: str)-> str`, a property could be that the returned value should always include `b` for any provided arguments.

Hypothesis helps us define those invariants and then takes care of generating as many tests as possible that assert that those properties always hold true. Usually, this is done by generating tests based on the domain of function arguments and ensuring that the properties hold true for all values. Obviously, testing all possible values would be too cumbersome, or even not doable at all since, for example, the values of the `str` domain are infinite. For this reason, Hypothesis is smart enough to know which values most frequently cause problems in a domain and will restrict the tests to those, also remembering which values caused problems to our code in the past, so that our test suite remains fast but also effective.

The most common usage of the Hypothesis testing library is as a replacement of the `pytest.mark.parametrize` decorator to automatically generate tests that run for different kinds of values based on the types of arguments.

In the case of our contacts book application, we might want to ensure that the contact book works for any kind of name the contacts have. We don't know whether our users will be from the USA, Europe, the Middle-East, or Asia, and so might have totally different concepts of names.

Using `pytest.mark.parametrize`, we could write a test that does that for some cases that come to mind:

```
import pytest
from contacts import Application

@pytest.mark.parametrize("name",
    ["Mario Alberto Rossi", "Étienne de La Boétie", "الزورق"]
)
def test_adding_contacts(name):
 app = Application()

 app.run(f"contacts add {name} 3456789")
 assert app._contacts == [(name, "3456789")]
```

The test will pass, and will try for some names and cases that come to mind:

```
$ pytest -v
================= test session starts =================
platform linux -- Python 3.8.6, pytest-6.0.2, py-1.9.0, pluggy-0.13.1 --
...
collected 3 items

tests/test_properties.py::test_adding_contacts[Mario Alberto Rossi] PASSED
[ 33%]
tests/test_properties.py::test_adding_contacts[\xc3\x89tienne de La
Bo\xc3\xa9tie] PASSED [ 66%]
tests/test_properties.py::test_adding_contacts[\u0627\u0644\u0632\u0648\u06
31\u0642] PASSED [100%]

================= 3 passed in 0.04s =================
```

But is this actually a good enough test? Let's see what happens if, instead of picking the values ourselves, we use Hypothesis to generate those tests. Implementing this change is as easy as replacing the `parametrize` decorator with a `hypothesis.given` decorator:

```
import hypothesis
import hypothesis.strategies as st

from contacts import Application


@hypothesis.given(st.text())
def test_adding_contacts(name):
    app = Application()

    app.run(f"contacts add {name} 3456789")
    assert app._contacts == [(name, "3456789")]
```

Now, running the Hypothesis version of the test leads to a much more interesting result compared to the version based on @parametrize; the Hypothesis-based version of the test actually fails:

```
$ pytest -v
================== test session starts ==================
platform linux -- Python 3.8.6, pytest-6.0.2, py-1.9.0, pluggy-0.13.1 --
...
collected 1 item

tests/test_properties.py::test_adding_contacts FAILED [100%]

===================== FAILURES =====================
_____ test_adding_contacts _____

    @given(st.text())
>   def test_adding_contacts(name):

tests/test_properties.py:8:
_ _ _ _ _ _ _ _ _ _ _ _ _ _ _ _ _ _ _ _ _ _ _ _ _ _ _
tests/test_properties.py:11: in test_adding_contacts
    app.run(f"contacts add {name} 3456789")
_ _ _ _ _ _ _ _ _ _ _ _ _ _ _ _ _ _ _ _ _ _ _ _ _ _ _
self = <contacts.Application object at 0x7f9a71fce850>,
text = 'contacts add 3456789'

    def run(self, text):
        ...

        if cmd == "add":
>           name, num = args.rsplit(maxsplit=1)
E           ValueError: not enough values to unpack (expected 2, got 1)

src/contacts/__init__.py:48: ValueError
-------------------- Hypothesis --------------------
Falsifying example: test_adding_contacts(
    name='',
)
================== 1 failed in 0.10s ==================
```

So, Hypothesis actually found a real bug in our software. If we don't provide a name at all, the software, instead of providing an error message, just crashes. We can see that Hypothesis tells us that the example that failed is the one where `name=''` and PyTest confirms that the string that was executed as a command was `text = 'contacts add 3456789'`. The line that crashed is the one that splits the name and number out of the `add` command arguments, so we have to handle the case where we can't split them apart because we only have one argument.

To do so, we can go back to the `Application.run` method and trap the exception that can come out of `args.rsplit`:

```
if cmd == "add":
    try:
        name, num = args.rsplit(maxsplit=1)
    except ValueError:
        print("A contact must provide a name and phone number")
        return
    try:
        self.add(name, num)
    except ValueError as err:
        print(err)
        return
```

Now, if we rerun our test, we will get a slightly different kind of failure, a failure in the test itself:

```
$ pytest -v
================= test session starts =================
platform linux -- Python 3.8.6, pytest-6.0.2, py-1.9.0, pluggy-0.13.1 --
...
collected 1 item

tests/test_properties.py::test_adding_contacts FAILED [100%]

===================== FAILURES =====================
_____ test_adding_contacts _____

    @given(st.text())
>   def test_adding_contacts(name):

tests/test_properties.py:8:
_ _ _ _ _ _ _ _ _ _ _ _ _ _ _ _ _ _ _ _ _ _ _ _
tests/test_properties.py:11: in test_adding_contacts
    app.run(f"contacts add {name} 3456789")
_ _ _ _ _ _ _ _ _ _ _ _ _ _ _ _ _ _ _ _ _ _ _ _
name = ''
```

```
       @given(st.text())
       def test_adding_contacts(name):
           app = Application()
           app.run(f"contacts add {name} 3456789")
   >       assert app._contacts == [(name, "3456789")]
   E       AssertionError: assert [] == [('', '3456789')]
   E         Right contains one more item: ('', '3456789')
   E         Full diff:
   E         - [('', '3456789')]
   E         + []

   tests/test_properties.py:15: AssertionError
   ---------------- Captured stdout call ----------------
   A contact must provide a name and phone number
   -------------------- Hypothesis --------------------
   Falsifying example: test_adding_contacts(
       name='',
   )
   ================= 1 failed in 0.10s =================
```

From `Captured stdout`, we can see that the error we emit when no name is provided was
properly reported, but our test failed because the assertion expects that a new contact is
always inserted while, in the case of a missing name, no contact gets added to our contact
book. So, in this case, Hypothesis found that our test itself is actually incomplete.

What we have to do is to change the assertion to ensure that the contact book actually
contains what we really expect in the case where no name is provided. In case there is no
name, the contact book should just be empty:

```
@given(st.text())
def test_adding_contacts(name):
    app = Application()

    app.run(f"contacts add {name} 3456789")

    name = name.strip()
    if name:
        assert app._contacts == [(name, "3456789")]
    else:
        assert app._contacts == []
```

At this point, rerunning the test will actually confirm that everything works as expected:

```
$ pytest -v
================= test session starts =================
platform linux -- Python 3.8.6, pytest-6.0.2, py-1.9.0, pluggy-0.13.1 --
...
```

```
collected 1 item

tests/test_properties.py::test_adding_contacts PASSED [100%]

================= 1 passed in 0.42s =================
```

We have seen how Hypothesis can help us to identify bugs and design tests, but it can actually do much more. It can even go as far as generating some tests for the most common properties for us.

Generating tests for common properties

Through the `hypothesis write` command, we can use Hypothesis to generate tests for use based on some of the most common properties functions might have. For example, if we want to ensure that the Python `sorted` method is idempotent and calling it twice leads to the exact same result, we can use `hypothesis write --idempotent sorted` to generate a test that verifies such a property:

```
$ hypothesis write --idempotent sorted

from hypothesis import given, strategies as st

@given(
    iterable=st.one_of(st.iterables(st.integers()),
st.iterables(st.text())),
    key=st.none(),
    reverse=st.booleans(),
)
def test_idempotent_sorted(iterable, key, reverse):
    result = sorted(iterable, key=key, reverse=reverse)
    repeat = sorted(result, key=key, reverse=reverse)
    assert result == repeat, (result, repeat)
```

Or, we could test that a pair of encode/decode functions leads back to the original result when chained using the `hypothesis write --roundtrip` generator.

If we want to check that for `json.loads` and `json.dumps`, for example, we could use `hypothesis write --roundtrip json.dumps json.loads`, which would generate the following code block:

```
$ hypothesis write --roundtrip json.dumps json.loads

import json
from hypothesis import given, strategies as st
```

```
@given(
    allow_nan=st.booleans(),
    check_circular=st.booleans(),
    cls=st.none(),
    default=st.none(),
    ensure_ascii=st.booleans(),
    indent=st.none(),
    obj=st.nothing(),
    object_hook=st.none(),
    object_pairs_hook=st.none(),
    parse_constant=st.none(),
    parse_float=st.none(),
    parse_int=st.none(),
    separators=st.none(),
    skipkeys=st.booleans(),
    sort_keys=st.booleans(),
)
def test_roundtrip_dumps_loads(
    allow_nan,
    check_circular,
    cls,
    default,
    ensure_ascii,
    indent,
    obj,
    object_hook,
    object_pairs_hook,
    parse_constant,
    parse_float,
    parse_int,
    separators,
    skipkeys,
    sort_keys,
):
    value0 = json.dumps(
        obj=obj,
        skipkeys=skipkeys,
        ensure_ascii=ensure_ascii,
        check_circular=check_circular,
        allow_nan=allow_nan,
        cls=cls,
        indent=indent,
        separators=separators,
        default=default,
        sort_keys=sort_keys,
    )
    value1 = json.loads(
        s=value0,
```

```
            cls=cls,
            object_hook=object_hook,
            parse_float=parse_float,
            parse_int=parse_int,
            parse_constant=parse_constant,
            object_pairs_hook=object_pairs_hook,
        )
        assert obj == value1, (obj, value1)
```

When refactoring code, implementing performance optimizations, or modifying code to port it from prior versions of Python, an essential property of the new implementation we are going to write is that it must retain the exact same behavior of the old implementation. The `hypothesis write --equivalent` command is able to do precisely this.

If, for example, we had two helper functions in `contacts/utils.py`, both meant to sum two numbers, as follows:

```
def sum1(a: int, b: int) -> int:
    return a + b

def sum2(a: int, b: int) -> int:
    return sum((a, b))
```

In that case, `hypothesis` could generate a test that verifies the fact that both functions lead to the exact same results:

```
$ hypothesis write --equivalent contacts.utils.sum1 contacts.utils.sum2
import contacts.utils
from hypothesis import given, strategies as st


@given(a=st.integers(), b=st.integers())
def test_equivalent_sum1_sum2(a, b):
    result_sum1 = contacts.utils.sum1(a=a, b=b)
    result_sum2 = contacts.utils.sum2(a=a, b=b)
    assert result_sum1 == result_sum2, (result_sum1, result_sum2)
```

While most of those tests could be written manually using `hypothesis.given`, it can be convenient having Hypothesis inspect the functions for you and pick the right types. Especially if you already did the effort of providing type hints for your functions, Hypothesis will usually be able to do the right thing.

To know all the generators that are available in your version of Hypothesis, you can run `hypothesis write --help`.

Summary

In this chapter, we saw how to have tested documentation that can guarantee user guides in sync with our code, and we saw how to make sure that our tests cover limit and corner cases we might not have considered through property-based testing.

Hypothesis can take away from you a lot of the effort of providing all possible values to a parameterized test, thereby making writing effective tests much faster, while `doctest` can ensure that the examples we write in our user guides remain effective in the long term, detecting whether any of them need to be updated when our code changes.

In the next chapter, we are going to shift our attention to the web development world, where we will see how to test web applications both from the point of view of functional tests and end-to-end tests.

Section 3: Testing for the Web 3

In this section, we will learn how to test web applications, web services, and APIs with Python, PyTest, and the most common testing tools available for WSGI frameworks.

This section comprises the following chapters:

- Chapter 11, *Testing for the Web: WSGI versus HTTP*
- Chapter 12, *End-to-End Testing with the Robot Framework*

Testing for the Web: WSGI versus HTTP

11

In the previous chapter, we saw how to test documentation and implement more advanced testing techniques in our test suites, such as property-based testing.

One of the primary use cases for Python has become web development. Python has many very effective and powerful web development frameworks. The most famous one is surely the Django web framework, but many more of them exist, including the Flask framework, the Pyramid framework, TurboGears2, and more. Each web framework has its own peculiarities and unique features that make it easy to build most of the different kinds of web applications using Python itself, but all of them share the same need of having to verify that the applications you built work properly and are tested. Thus in this chapter, we are going to see how we can test HTTP-based applications on both the client and server side, how we can do that using pytest, and how the techniques presented differ from framework-specific tests.

In this chapter, we will cover the following topics:

- Testing HTTP
- Testing WSGI with WebTest
- Using WebTest with web frameworks
- Writing Django tests with Django's test client

In this chapter, we are going to reverse the approach a bit and we are going to violate the **Test-Driven Development (TDD)** principle by implementing the code first and introducing tests for it after. The reason for this is that by introducing the system under test first we can illustrate more clearly some details of the tests. If you already know how the tested software works, it's easier to understand why the tests do the things they do, so for the purposes of this chapter we will briefly abandon our best practices and focus on the code first, and the tests after.

Technical requirements

We need a working Python interpreter with pytest, but for some sections in this chapter, we will also have to install other libraries and frameworks. As usual, all of them can be installed with `pip`:

```
$ pip install pytest
```

For the *Testing HTTP* section, we are going to need the `requests` library and the `requests-mock` testing library:

```
$ pip install requests requests-mock
```

For the *Testing WSGI with WebTest* section, we are going to need `webtest`:

```
$ pip install webtest
```

And for the paragraphs regarding testing web frameworks, we are going to need the targeted web frameworks installed, even though you aren't going to use all of them concurrently in a real project:

```
$ pip install flask django pyramid turbogears2
```

The examples have been written on Python 3.7, pytest 6.0.2, Requests 2.24.0, Requests-Mock 1.8.0, WebTest 2.0.35, Django 3.1.4, Flask 1.1.2, Pyramid 1.10.5, and TurboGears 2.4.3, but should work on most modern Python versions. You can find the code files present in this chapter on GitHub at `https://github.com/PacktPublishing/Crafting-Test-Driven-Software-with-Python/tree/main/Chapter11`.

Testing HTTP

A frequent need when working with networking based applications is that we have to test both the server and client. If we are writing a distributed application, we are probably going to write both the client and the server ourselves, and that means we'll want to test both of them just as we did with our Chat application in previous chapters.

While we might want to have a limited number of tests that connect to a real running server, that quickly becomes too expensive if we involve real networking, and could also result in errors related to the maximum amount of open connections our system can handle, along with the time it takes to actually shut down those connections.

So we need to be able to test the client side of the application without having to connect to a real server for the majority of our tests, or our test suite will quickly become unmaintainable.

Let's suppose we are writing a very simple `httpclient` command-line application that will allow us to request any URL that we want with the most common HTTP methods:

```
$ python -m httpclient GET http://www.amazon.com/
<!DOCTYPE html>
<html class="a-no-js" lang="en-us">
    <head>
        <title dir="ltr">Amazon.com</title>
        ...
```

To do so, we would first need a class able to perform HTTP requests, which we are going to call just `HTTPClient`. Our `HTTPClient` exposes support for `GET`, `POST`, and `DELETE` (we could easily expose more, but for the sake of simplicity we will limit our client to those most common methods), and a `follow` method that allows us to access nested paths relative to the current URL.

To implement this object we are going to rely on the `requests` library for most of the heavy lifting of HTTP processing, thus we can run `import requests` and rely on it for most of our methods' implementations. Let's create a `src/httpclient/__init__.py` file where we can place our `HTTPClient` object:

```python
import urllib.parse
import requests


class HTTPClient:
    def __init__(self, url):
        self._url = url

    def GET(self):
        return requests.get(self._url).text

    def POST(self, **kwargs):
        return requests.post(self._url, data=kwargs).text

    def DELETE(self):
        return requests.delete(self._url).text
```

```
def follow(self, path):
    baseurl = self._url
    if not baseurl.endswith("/"):
        baseurl += "/"
    return HTTPClient(urllib.parse.urljoin(baseurl, path))
```

The only method not directly relying on `requests` is `HTTPClient.follow`, which uses the `urllib.parse` standard library module to navigate the URL tree.

Given the current URL, which is used as the base, the method is going to return a new `HTTPClient` that points to a path nested within the same URL. For example, if we have a client pointing to `"http://www.google.com/"`, then using `HTTPClient.follow("me")` would give us back a new client instance through which we can request `http://www.google.com/me`.

 Notice that this is a very naïve implementation that takes for granted the fact that the base URL doesn't have any parameters. A more robust implementation could be achieved if we actually parsed the URL and encoded it back into a string, so that we can isolate the path from the rest of the URL.

Now that we have the client in place, the remaining parts are those involved in exposing it on the command line, so that we can use the `python -m httpclient` command to perform HTTP requests.

The first piece we need to do so is the `parse_args` function. This function will be in charge of taking arguments from the command line (thus from `sys.argv`) and converting them to the options for `HTTPClient`:

```
import sys

def parse_args():
    cmd = sys.argv[0]
    args = sys.argv[1:]
    try:
        method, url, *params = args
    except ValueError:
        raise ValueError("Not enough arguments, "
                        "at least METHOD URL must be provided")

    try:
        params = dict((p.split("=", 1) for p in params))
    except ValueError:
        raise ValueError("Invalid request body parameters. "
                        "They must be in name=value format, "
```

```
                          f"not {params}")

        return method.upper(), url, params
```

The first code block is just going to separate the HTTP method, the URL we want to request, and the various params we want to provide it. The HTTP method accepts any number of params, so we could have zero or many.

The second code block is meant to parse `params` from a `"name=value"` format to a dictionary we can pass to the `HTTPClient.POST` method.

Finally, the function returns the `HTTPClient` method we have to invoke (`GET`, `POST`, or `DELETE`), the URL for which we have to invoke it, and the `params` dictionary containing all parameters.

Those three values are useful to the real `main` function of our application to properly use the `HTTPClient` object. So the next step is to implement this `main` function so that we can invoke it from the command line:

```
def main():
    try:
        method, url, params = parse_args()
    except ValueError as err:
        print(err)
        return

    client = HTTPClient(url)
    print(getattr(client, method)(**params))
```

`main` invokes `parse_args`, creates a `client` object, and then invokes the method requested by `parse_args` on it and prints the returned value.

The remaining pieces we need to handle are, firstly, to create a `src/httpclient/__main__.py` file where we invoke the `main` function:

```
from httpclient import main

main()
```

And then a `src/setup.py` file that allows us to install the package and invoke it from the command line:

```
from setuptools import setup

setup(name='httpclient', packages=['httpclient'])
```

If everything worked as expected, installing our package should allow us to invoke it from the command line to perform HTTP requests:

```
$ pip install -e ./src
Obtaining file://./src
Installing collected packages: httpclient
...
Successfully installed httpclient

$ python -m httpclient GET http://httpbin.org/get
{
  "args": {},
  "headers": {
    "Accept": "*/*",
    "Accept-Encoding": "gzip, deflate",
    "Host": "httpbin.org",
    "User-Agent": "python-requests/2.24.0",
  },
  "url": "http://httpbin.org/get"
}
```

Now that all the pieces are in place, we can move on to see how to test the `HTTPClient` object.

Testing HTTP clients

If we had to test our `HTTPClient`, we would have to perform HTTP requests through those methods to confirm they actually do what we want. To do so, we could use `httpbin.org`, which is a service that accepts any kind of request and echoes back what was submitted. This would allow us to verify that we are submitting what we expected we would send to the server:

```
import json
from httpclient import HTTPClient


class TestHTTPClient:
    def test_GET(self):
```

```
        client = HTTPClient(url="http://httpbin.org/get")
        response = client.GET()

        assert '"Host": "httpbin.org"' in response
        assert '"args": {}' in response

    def test_GET_params(self):
        client = HTTPClient(url="http://httpbin.org/get?alpha=1")
        response = client.GET()
        response = json.loads(response)

        assert response["headers"]["Host"] == "httpbin.org"
        assert response["args"] == {"alpha": "1"}

    def test_POST(self):
        client = HTTPClient(url="http://httpbin.org/post?alpha=1")
        response = client.POST(beta=2)
        response = json.loads(response)

        assert response["headers"]["Host"] == "httpbin.org"
        assert response["args"] == {"alpha": "1"}
        assert response["form"] == {"beta": "2"}

    def test_DELETE(self):
        client = HTTPClient(url="http://httpbin.org/anything/27")
        response = client.DELETE()

        assert '"method": "DELETE"' in response
        assert '"url": "http://httpbin.org/anything/27"' in response

    def test_follow(self):
        client = HTTPClient(url="http://httpbin.org/anything")

        assert client._url == "http://httpbin.org/anything"

        client2 = client.follow("me")

        assert client2._url == "http://httpbin.org/anything/me"
```

Saving those tests as `tests/test_httpclient.py` will provide us with a running test suite that confirms that `HTTPClient` works as expected. The problem is that running the tests with this approach can take a while. Running just a few simple tests already takes more than a second to run:

```
$ pytest -v -s
====================== test session starts ======================
platform linux -- Python 3.8.6, pytest-6.0.2, py-1.9.0, pluggy-0.13.1
...
```

```
collected 5 items

tests/test_httpclient.py::TestHTTPClient::test_GET PASSED
tests/test_httpclient.py::TestHTTPClient::test_GET_params PASSED
tests/test_httpclient.py::TestHTTPClient::test_POST PASSED
tests/test_httpclient.py::TestHTTPClient::test_DELETE PASSED
tests/test_httpclient.py::TestHTTPClient::test_follow PASSED
======================= 5 passed in 1.37s =======================
```

Also, the tests might randomly fail due to network issues or errors on the remote server, so they could easily become flaky. Slow and flaky tests are something we must avoid in a test suite, so this approach of involving real networking is not something we can rely on in our test suite.

The solution to both those problems is to replace the remote server, and thus the need for networking, with a fake implementation. In our specific case, as we used the `requests` library to perform HTTP requests to the server, we can prepare ready-made answers for our requests using the `requests-mock` library, which allows us to mock requests by replacing them with pre-baked responses.

To replace our real requests with fake ones, we just have to wrap them in a `requests_mock.Mocker()` context manager, which comes from the `requests_mock` module made available by the `requests-mock` library. Once we have the mocker object, we can use it to drive what has to be mocked (which URL, method, and so on) and serve ready-made answers for all the requests that match those filters.

For example, to mock a GET request, we could create the `HTTPClient` and before invoking `client.GET` we could wrap that method with the `Mocker` and thus set up a ready-made answer for any GET request against the same URL as the client one:

```
client = HTTPClient(url="http://httpbin.org/get")

with requests_mock.Mocker() as m:
    m.get(client._url, text='{"Host": "httpbin.org", "args": {}}')
    response = client.GET()
```

The `text`, `json`, and `content` arguments of the mocker can be used to provide the response (as text, JSON, or binary) we want to serve back when the URL is requested with the specified method. In this case, for example, we provided the response in text format even though it contains a JSON string. In the following examples, we are going to use the `json` argument, so that we can see both of them in action.

Now we can adapt all our tests to use `requests_mock` so that they no longer have to take a networking roundtrip to pass:

```python
import json
from httpclient import HTTPClient
import requests_mock


class TestHTTPClient:
    def test_GET(self):
        client = HTTPClient(url="http://httpbin.org/get")
        with requests_mock.Mocker() as m:
            m.get(client._url,
                    text='{"Host": "httpbin.org", "args": {}}')
            response = client.GET()

        assert '"Host": "httpbin.org"' in response
        assert '"args": {}' in response

    def test_GET_params(self):
        client = HTTPClient(url="http://httpbin.org/get?alpha=1")
        with requests_mock.Mocker() as m:
            m.get(client._url,
                    text='''{"headers": {"Host": "httpbin.org"},
                            "args": {"alpha": "1"}}''')
            response = client.GET()

        response = json.loads(response)
        assert response["headers"]["Host"] == "httpbin.org"
        assert response["args"] == {"alpha": "1"}

    def test_POST(self):
        client = HTTPClient(url="http://httpbin.org/post?alpha=1")
        with requests_mock.Mocker() as m:
            m.post(client._url, json={"headers": {"Host": "httpbin.org"},
                                      "args": {"alpha": "1"},
                                      "form": {"beta": "2"}})
            response = client.POST(beta=2)

        response = json.loads(response)
        assert response["headers"]["Host"] == "httpbin.org"
        assert response["args"] == {"alpha": "1"}
        assert response["form"] == {"beta": "2"}

    def test_DELETE(self):
        client = HTTPClient(url="http://httpbin.org/anything/27")
        with requests_mock.Mocker() as m:
            m.delete(client._url, json={
```

```
                    "method": "DELETE",
                    "url": "http://httpbin.org/anything/27"
             })
             response = client.DELETE()

         assert '"method": "DELETE"' in response
         assert '"url": "http://httpbin.org/anything/27"' in response

    def test_follow(self):
        ...
```

The `test_follow` test remains unchanged as it didn't involve any networking, while the other tests are now wrapped in a `requests_mock.Mocker()` surrounding the `client.GET`, `client.POST` and `client.DELETE` calls.

With those changes, the impact on our test suite is immediately visible. Tests that previously took more than a second to run now take just a few milliseconds:

```
$ pytest -v -s
======================= test session starts =======================
platform linux -- Python 3.8.6, pytest-6.0.2, py-1.9.0, pluggy-0.13.1
...
collected 5 items

tests/test_httpclient.py::TestHTTPClient::test_GET PASSED
tests/test_httpclient.py::TestHTTPClient::test_GET_params PASSED
tests/test_httpclient.py::TestHTTPClient::test_POST PASSED
tests/test_httpclient.py::TestHTTPClient::test_DELETE PASSED
tests/test_httpclient.py::TestHTTPClient::test_follow PASSED
======================= 5 passed in 0.03s =======================
```

While this approach is fast, robust, and allows us to test that the client is properly able to process and react to answers, it doesn't really test that the client and the server are able to work together. Yes, we know that the client behaves like we meant it to behave, but it doesn't in any way guarantee that once we put it in front of a real server, the two will speak the same language.

If the server changes its responses in a way that differs from the one we hardcoded in our tests, we will notice that the client doesn't work anymore with our server.

To address this limitation without having to involve a real networking layer, we are going to see how we can write integration tests using **WebTest** and the **WSGI** protocol.

Testing WSGI with WebTest

While we have seen how to test client without connecting to a real server, we can't rely only on faked messages to confirm that our application works. If we are going to change server responses, the tests wouldn't even notice and would continue to pass while in reality, the client has stopped working. How can we detect those kinds of issues without involving real networking? The **WSGI** (**Web Server Gateway Interface**) protocol and WebTest library come in hand to do exactly that, set up a client-server communication that involves no networking at all.

When we create web applications in Python, the most frequent way they work is through an **application server**. The application server will be the one receiving HTTP requests, decoding them, and forwarding them to the real web application. Forwarding those requests to the web application and receiving back responses via the WSGI protocol is usually the communication channel of choice for Python.

The WSGI protocol is a pure Python protocol, thus relies solely on being able to invoke a Python function passing some specific arguments. All the communication in WSGI happens in-memory and involves no dedicated parsing, and thus is very fast and usually suitable for integration in web applications. A complete description of WSGI is available in PEP 333 (`https://www.python.org/dev/peps/pep-3333/`).

The most basic WSGI application is a simple `callable` (a function, method, or function object) that accepts two arguments (`environ` and `start_response`) and responds with an `iterable` containing the output to be sent back to the client after having invoked `start_response` to set up the response headers.

So the basic *"Hello World"* kind of application in WSGI would look as follows:

```python
class Application:
    def __call__(self, environ, start_response):
        start_response(
            '200 OK',
            [('Content-type', 'text/plain; charset=utf-8')]
        )
        return ["Hello World".encode("utf-8")]
```

The `environ` argument will contain all information about the environment within which our request is being processed, including information about the request itself, such as `REQUEST_METHOD`, `HTTP_HOST`, `PATH_INFO`, `QUERY_STRING`, and many more values. `start_response` is a function we can invoke to tell the application server that we are ready to send back our response and inform it about the response type and the HTTP headers that have to be sent back.

In our case, for every request, we always send back an HTTP 200 response informing the client that we are going to send some text encoded in UTF-8 by providing a Content-Type header.

Then we return the iterable containing the response, which in this case is a list containing the "Hello World" string encoded as UTF-8 as specified in our Content-Type.

Now that we have our WSGI application, we can save it in the src/wsgiwebtest/__init__.py file and move forward to see how we can attach it to the application server.

For the sake of this example, we are going to use a very basic application server provided by the Python standard library itself in the wsgiref module, simple_server.WSGIServer. To be able to start our application we are going to create a src/wsgiwebtest/__main__.py file where we are going to place a main function that creates the WSGIServer and attaches it to our web application:

```python
from wsgiref.simple_server import make_server
from wsgiwebtest import Application

def main():
    app = Application()
    with make_server('', 8000, app) as httpd:
        print("Serving on port 8000...")
        httpd.serve_forever()

main()
```

All our main has to do is to create the Application object and pass it to the make_server function, which will create an application server for that application. Once the server is available we can start serving requests through the httpd.server_forever method.

The last step before we can actually try our "*Hello World* application is to create a setup.py file so that we can install our package. So let's save a basic one as src/setup.py, containing the following:

```python
from setuptools import setup

setup(name='wsgiwebtest', packages=['wsgiwebtest'])
```

Now that we have all the pieces in place, we can install our application and start it:

```
$ pip install -e ./src
Obtaining file:///./src
Installing collected packages: wsgiwebtest
```

```
...
Successfully installed wsgiwebtest

$ python -m wsgiwebtest
Serving on port 8000...
```

Pointing our browser to `http://localhost:8000/` should greet us with a simple **Hello World** phrase:

Figure 11.1 – Hello World answer from our WSGI application

Now that we have a working web application, we want to evolve it to make it a bit more interesting. We are going to turn it into a simple clone of `httpbin.org`. To do so we are going to use the same exact tests we wrote for our `HTTPClient` package, port them to use **WebTest**, and use them to drive the development of our WSGI application.

The first step is to take our existing `TestHTTPClient.test_GET` test and port it to use `webtest` to verify our web application, saving it as `tests/test_wsgiapp.py`:

```python
import webtest

from wsgiwebtest import Application


class TestWSGIApp:
    def test_GET(self):
        client = webtest.TestApp(Application())
        response = client.get("http://httpbin.org/get").text

        assert '"Host": "httpbin.org"' in response
        assert '"args": {}' in response
```

The main difference is that instead of building an `HTTPClient` instance, we build a `webtest.TestApp` for the application we want to test, which in this case is `wsgiwebtest.Application`. Then we ask `TestApp` to perform a GET request against a specific URL using the `TestApp.get` method. While we can specify a complete URL including the domain, it won't matter too much, as `TestApp` will always direct the request to the application under test, so even though here we wrote `"http://httpbin.org"`, in reality the request won't go to `"httpbin.org"` but to `wsgiwebtest.Application`. This allows us to test our application by simulating whatever domain we want to serve it on.

Then the response returned by this request can be decoded as text or JSON as we did for the `requests` library, using the `.text` or `.json` properties. In this case, we are going to retain the existing test behavior and decode it as text even though the response is actually in JSON format.

Running the test will obviously fail because right now our web application only responds with `"Hello World"` to every request, but it proves that our test actually reached our web application and got back the `"Hello World"` response:

```
_____ TestWSGIApp.test_GET _____

self = <test_wsgiapp.TestWSGIApp object at 0x7fc6d64feaf0>

    def test_GET(self):
        client = webtest.TestApp(Application())
        response = client.get("http://httpbin.org/get")
>       assert '"Host": "httpbin.org"' in response
E       assert '"Host": "httpbin.org"' in <200 OK text/plain body=b'Hello World'>
```

Now that we know that `webtest` is actually working as expected and is doing the GET request against our web application, let's start porting all our other tests to use `webtest`. The approach is nearly the same for all of them. Instead of building an `HTTPClient` instance, we are going to build a `webtest.TestApp` and use its `.get`, `.post`, and `.delete` methods to perform the requests:

```python
import webtest

from wsgiwebtest import Application


class TestWSGIApp:
    def test_GET(self):
        client = webtest.TestApp(Application())
        response = client.get("http://httpbin.org/get").text

        assert '"Host": "httpbin.org"' in response
```

```
        assert '"args": {}' in response

    def test_GET_params(self):
        client = webtest.TestApp(Application())

        response = client.get(url="http://httpbin.org/get?alpha=1").json

        assert response["headers"]["Host"] == "httpbin.org"
        assert response["args"] == {"alpha": "1"}

    def test_POST(self):
        client = webtest.TestApp(Application())

        response = client.post(url="http://httpbin.org/get?alpha=1",
                               params={"beta": "2"}).json

        assert response["headers"]["Host"] == "httpbin.org"
        assert response["args"] == {"alpha": "1"}
        assert response["form"] == {"beta": "2"}

    def test_DELETE(self):
        client = webtest.TestApp(Application())

        response = client.delete(url="http://httpbin.org/anything/27").text
        assert '"method": "DELETE"' in response
        assert '"url": "http://httpbin.org/anything/27"' in response
```

The assertion part of the tests remained unmodified from the original tests we copied, the only part that slightly changed is how we perform the requests.

Like the first test, those new tests will currently all fail because our web application will always respond with `"Hello World"` to all of them. So the next step is to change our web application to make it respond as the tests expect.

We can open our existing `src/wsgiwebtest/__init__.py` file and tweak the `Application.__call__` method to make it recognize the requested host, URL, and method while also parsing the received request parameters from both the URL and the request body:

```
    import urllib.parse


    class Application:
        def __call__(self, environ, start_response):
            start_response(
                '200 OK',
                [('Content-type', 'application/json; charset=utf-8')]
```

```
    )

    form_params = {}
    if environ.get('CONTENT_TYPE') ==
            'application/x-www-form-urlencoded':
        req_body = environ["wsgi.input"].read().decode("ascii")
        form_params = {
            k: v for k, v in urllib.parse.parse_qsl(req_body)
        }

    if environ.get("SERVER_PORT") == "80":
        host = environ["SERVER_NAME"]
    else:
        host = environ["HTTP_HOST"]

    return [json.dumps({
        "method": environ["REQUEST_METHOD"],
        "headers": {"Host": host},
        "url": "{e[wsgi.url_scheme]}://{host}{e[PATH_INFO]}".format(
            e=environ,
            host=host
        ),
        "args": {
            k: v for k, v in
                urllib.parse.parse_qsl(environ["QUERY_STRING"])
        },
        "form": form_params
}).encode("utf-8")]
```

The `start_response` invocation is nearly the same, we just changed the reported `Content-Type` to be `application/json` instead of `text/plain` as we are going to serve back a JSON response.

Right after this, `form_params` is meant to contain all the parameters provided through the request body. If what we received is a POST request, it's probably going to have a request body where the majority of the parameters are provided. The request body could provide those parameters encoded in various ways, but as it's the simplest one (and the one our tests used), we are going to support only the `"application/x-www-form-urlencoded"` encoding. So if the request we received has that content type, we will also parse the request body (coming from `environ["wsgi.input"]`) and extract the parameters from there.

The subsequent code block that initializes the `host` variable is instead meant to find the host and port from which the request came, so that we can send it back into the `Host` field of the `headers` dictionary in our response as the tests expect. The test expects that if the request is targeted to the standard HTTP port, `80`, the port is omitted in the returned host. So we are going to only report the port when it's not `80` and we are going to limit ourselves to the `SERVER_NAME` when the port is `80`.

The last block is actually focused on building back the response, so it uses `json.dumps` to encode a dictionary with all the data as text. The dictionary is going to contain the fields our tests care about, meaning `method` and `headers.Host` for the HTTP method that was used to perform the request, and the `Host` against which the request was targeted (in our tests, this is `httpbin.org`). This will also contain the `args` key for all the parameters provided in the query string, and thus in the URL itself, while separating the parameters that were provided in the request body in the `form` key. Finally, the `url` key contains the fully qualified URL that was requested.

This should guarantee a behavior very similar to the one that the real `httpbin.org` provides, albeit heavily simplified. Saving back our new code and trying to rerun the tests should prove that we implemented something that is similar enough to make our tests pass:

```
$ pytest -v -s
==================== test session starts ====================
platform linux -- Python 3.8.6, pytest-6.0.2, py-1.9.0, pluggy-0.13.1
...
collected 4 items

tests/test_wsgiapp.py::TestHTTPClient::test_GET PASSED [ 25%]
tests/test_wsgiapp.py::TestHTTPClient::test_GET_params PASSED [ 50%]
tests/test_wsgiapp.py::TestHTTPClient::test_POST PASSED [ 75%]
tests/test_wsgiapp.py::TestHTTPClient::test_DELETE PASSED [100%]
==================== 4 passed in 0.08s ====================
```

Our tests passed, proving that our web application is similar enough to the original one we meant to copy, and even though there is tons of space for improvement, it demonstrated how we can implement tests for web applications that don't need any networking at all. This is made clear by the fact that our tests using **WebTest** still complete in a matter of milliseconds, similar to the tests where we used `requests-mock`, while the tests that involved real networking took more than a second.

If we want to go even further, using a little bit of dependency injection, we could easily modify our HTTPClient object to work with both the requests module and the webtest.TestApp object, as they are similar enough that we could write end-to-end tests that go from HTTPClient down to wsgiwebtest.Application without ever involving any HTTP parsing or networking.

Going in this direction requires a brief change to our original HTTPClient to allow us to provide a replacement for the requests module at initialization time. By default, we are going to keep using the requests module, but anyone could pass a different object to HTTPClient.__init__ and replace it:

```
class HTTPClient:
    def __init__(self, url, requests=requests):
        self._url = url
        self._requests = requests

    def follow(self, path):
        baseurl = self._url
        if not baseurl.endswith("/"):
            baseurl += "/"
        return HTTPClient(urllib.parse.urljoin(baseurl, path))

    def GET(self):
        return self._requests.get(self._url).text

    def POST(self, **kwargs):
        return self._requests.post(self._url, kwargs).text

    def DELETE(self):
        return self._requests.delete(self._url).text
```

Then we have to use self._requests everywhere instead of just requests. The TestApp and requests interfaces are similar enough that the only change we actually need to the rest of the code is to omit the name of the argument (data=) from the post method and invoke it with a positional argument. This is because in requests, the argument is named data, while in TestApp it is named params. Passing it by position means that we don't need to worry about what name it has.

Now our `HTTPClient` is ready to accept a replacement for `requests` and we can take back the original version of the tests we wrote for `HTTPClient` (the one that didn't use `requests-mock`) and pass an instance of `webtest.TestApp(wsgiwebtest.Application)` as the replacement for `requests`:

```python
import json

import webtest

from wsgiwebtest import Application
from httpclient import HTTPClient


class TestHTTPClientWebTest:
    def test_GET(self):
        client = HTTPClient(url="http://httpbin.org/get",
                            requests=webtest.TestApp(Application()))
        response = client.GET()

        assert '"Host": "httpbin.org"' in response
        assert '"args": {}' in response

    def test_GET_params(self):
        client = HTTPClient(url="http://httpbin.org/get?alpha=1",
                            requests=webtest.TestApp(Application()))
        response = client.GET()
        response = json.loads(response)

        assert response["headers"]["Host"] == "httpbin.org"
        assert response["args"] == {"alpha": "1"}

    def test_POST(self):
        client = HTTPClient(url="http://httpbin.org/post?alpha=1",
                            requests=webtest.TestApp(Application()))
        response = client.POST(beta=2)
        response = json.loads(response)

        assert response["headers"]["Host"] == "httpbin.org"
        assert response["args"] == {"alpha": "1"}
        assert response["form"] == {"beta": "2"}

    def test_DELETE(self):
        client = HTTPClient(url="http://httpbin.org/anything/27",
                            requests=webtest.TestApp(Application()))
        response = client.DELETE()
```

```
        assert '"method": "DELETE"' in response
        assert '"url": "http://httpbin.org/anything/27"' in response
```

If we save those tests as `tests/test_client_webtest.py`, they will keep working exactly like before, but they will submit real requests to `wsgiwebtest.Application` through the WSGI protocol, thus making sure that both the server and the client are able to work together:

```
$ pytest -v -s
====================== test session starts ======================
platform linux -- Python 3.8.6, pytest-6.0.2, py-1.9.0, pluggy-0.13.1
...
collected 4 items

tests/test_client_webtest.py::TestHTTPClientWebTest::test_GET PASSED [ 25%]
tests/test_client_webtest.py::TestHTTPClientWebTest::test_GET_params PASSED
[ 50%]
tests/test_client_webtest.py::TestHTTPClientWebTest::test_POST PASSED [
75%]
tests/test_client_webtest.py::TestHTTPClientWebTest::test_DELETE PASSED
[100%]
====================== 4 passed in 0.08s ======================
```

Any change to one of the two that makes it incompatible with the other would immediately cause the tests to fail, thus verifying that the two work correctly together without any of the overhead of network-based communication or the flakiness that it involves.

All this is made possible by the fact that we used the WSGI standard to develop our web application and, as we are going to see in the next section, WSGI is the most widespread web development standard in Python and is supported by all major web frameworks.

Using WebTest with web frameworks

We have seen how to use WebTest with a plain WSGI application, but thanks to the fact that WSGI is widely adopted by all major web frameworks, it's possible to use WebTest with nearly all Python web frameworks.

To showcase how WebTest is able to work with most Python web frameworks, we are going to replicate our `httpbin` in four web frameworks: Django, Flask, Pyramid, and TurboGears2, and for all of them we are going to use the same exact test suite. So we will share a single test suite between four different frameworks.

The first step is to create a test suite that can verify that our web applications are starting correctly. We are going to do so by adding a test that verifies all four web applications' answering with a *"Hello World"* message on the index of the website.

The first step is to create a `tests/test_wsgiapp.py` file that's going to contain our only test for now:

```
import webtest


class TestWSGIApp:
    def test_home(self, wsgiapp):
        client = webtest.TestApp(wsgiapp)
        response = client.get("http://httpbin.org/").text

        assert 'Hello World' in response
```

The test is fairly simple – it takes a WSGI application and checks that, on the index of the website, the response contains the `"Hello World"` string.

The interesting part is how we are going to provide that `wsgiapp` object, as it has to be different for each web framework. So we are going to add an option to our test suite to choose which web framework to use and thus which application to create.

We are going to do so by creating a `tests/conftest.py` file that is going to contain both the new option and the fixture to create the `wsgiapp`. The first thing we want to add is support for the new option:

```
import pytest


def pytest_addoption(parser):
    parser.addoption(
        "--framework", action="store",
        help="Choose which framework to use for "
            "the web application: [tg2, django, flask, pyramid]"
    )
```

If things work correctly, once we save the `tests/conftest.py` file, running `pytest --help` will properly show the new option in the custom ones:

```
$ pytest --help
...
custom options:
  --framework=FRAMEWORK
                        Choose which framework to use for the
                        web application: [tg2, django, flask, pyramid]
```

Now that we have the option available, we must create the fixture that is going to use the option, the `wsgiapp` fixture. As it's a fixture available for all our test suites, we can just add it to the `conftest.py` file under the new option:

```
@pytest.fixture
def wsgiapp(request):
    framework = request.config.getoption("--framework")

    if framework == "tg2":
        from wbtframeworks.tg2 import make_application
    elif framework == "flask":
        from wbtframeworks.flask import make_application
    elif framework == "pyramid":
        from wbtframeworks.pyramid import make_application
    elif framework == "django":
        from wbtframeworks.django import make_application
    else:
        make_application = None

    if make_application is not None:
        return make_application()

    if framework is None:
        raise ValueError("Please pick a framework with --framework option")
    else:
        raise ValueError(f"Invalid framework {framework}")
```

The first thing that the fixture does is retrieve the selected framework through the option. Then, depending on which framework was selected, it's going to import the function that creates a new WSGI application from the module dedicated to that framework.

For convenience, we added all four modules (`tg2`, `flask`, `pyramid`, and `django`) under the same `wbtframeworks` package, which is the one we are going to install.

Once a framework is selected and the `make_application` function is imported, the fixture will just return the new application built by the factory function. The remaining lines of code are to handle the case where the user picks an unsupported framework (or no framework at all).

Running pytest now should lead to it correctly complaining that we have picked no framework:

```
$ pytest -v
================= test session starts =================
platform linux -- Python 3.8.6, pytest-6.0.2, py-1.9.0, pluggy-0.13.1
collected 1 item

tests/test_wsgiapp.py::TestWSGIApp::test_home ERROR [100%]

======================= ERRORS =======================
_____ ERROR at setup of TestWSGIApp.test_home _____

request = <SubRequest 'wsgiapp' for <Function test_home>>

    @pytest.fixture
    def wsgiapp(request):
        ...
        elif framework is None:
>           raise ValueError("Please pick a framework with --framework option")
E           ValueError: Please pick a framework with --framework option

tests/conftest.py:31: ValueError
=================== short test summary info ===================
ERROR tests/test_wsgiapp.py::TestWSGIApp::test_home -
        ValueError: Please pick a framework with --framework option
======================= 1 error in 0.15s =======================
```

To confirm that the option is working as expected, we can run pytest with the `--framework=flask` option to see what happens:

```
$ pytest -v --framework=flask
================= test session starts =================
platform linux -- Python 3.8.6, pytest-6.0.2, py-1.9.0, pluggy-0.13.1
collected 1 item

tests/test_wsgiapp.py::TestWSGIApp::test_home ERROR [100%]

======================= ERRORS =======================
_____ ERROR at setup of TestWSGIApp.test_home _____

request = <SubRequest 'wsgiapp' for <Function test_home>>
```

```
    @pytest.fixture
    def wsgiapp(request):
        framework = request.config.getoption("--framework")
        if framework == "tg2":
>           from wbtframeworks.tg2 import make_application
E           ModuleNotFoundError: No module named 'wbtframeworks'

tests/conftest.py:31: ValueError
==================== short test summary info ====================
ERROR tests/test_wsgiapp.py::TestWSGIApp::test_home -
        ModuleNotFoundError: No module named 'wbtframeworks'
======================= 1 error in 0.15s =======================
```

In this second case, it recognized the option correctly, but it complained that the wbtframeworks package is not yet installed. That's expected as we haven't yet even created it.

First, let's create a src/setup.py file to make the wbtframeworks package installable:

```
from setuptools import setup

setup(name='wbtframeworks', packages=['wbtframeworks'])
```

Now that the wbtframeworks package is installable, the next step is to create the package itself, by creating the src/wbtframeworks/__init__.py file and then installing it:

```
$ pip install -e src
Obtaining file://src
Installing collected packages: wbtframeworks
  Running setup.py develop for wbtframeworks
Successfully installed wbtframeworks
```

Now that the package is available and installed in editable mode, we have to create the structure for the four frameworks.

For the sake of keeping things short, as the sole purpose of those web applications is to showcase how the same test suite can work against the four of them, we are going to use all four frameworks in minimal mode, constraining the application to a single file.

The first one we are going to add is the src/wbtframeworks/flask/__init__.py file, to add support for Flask:

```
from flask import Flask
app = Flask(__name__)

@app.route('/')
def hello_world():
```

```
        return 'Hello World'


    def make_application():
        return app.wsgi_app
```

We can confirm this minimal application works as expected by running our tests with `pytest --frameworks=flask`:

```
$ pytest -v --framework=flask
====================== test session starts ======================
platform linux -- Python 3.8.6, pytest-6.0.2, py-1.9.0, pluggy-0.13.1
collected 1 item

tests/test_wsgiapp.py::TestWSGIApp::test_home PASSED [100%]

======================= 1 passed in 0.13s =======================
```

We use the same technique to create a `src/wbtframeworks/pyramid/__init__.py` file for the Pyramid application:

```
from pyramid.config import Configurator
from pyramid.response import Response


def hello_world(request):
    return Response('Hello World!')


def make_application():
    with Configurator() as config:
        config.add_route('hello', '/')
        config.add_view(hello_world, route_name='hello')
        return config.make_wsgi_app()
```

Likewise, let's create the `src/wbtframeworks/tg2/__init__.py` for the TurboGears2 application as follows:

```
from tg import expose, TGController
from tg import MinimalApplicationConfigurator


class RootController(TGController):
    @expose()
    def index(self):
        return 'Hello World'
```

```
def make_application():
    config = MinimalApplicationConfigurator()
    config.update_blueprint({
        'root_controller': RootController()
    })

    return config.make_wsgi_app()
```

And finally, create a `src/wbtframeworks/django/__init__.py` file for the Django application:

```
import sys
import os
from django.conf.urls import re_path
from django.conf import settings
from django.http import HttpResponse

settings.configure(
    DEBUG=True,
    ROOT_URLCONF=sys.modules[__name__],
    ALLOWED_HOSTS=["httpbin.org"]
)


def home(request):
    return HttpResponse('Hello World')

urlpatterns = [
    re_path(r'^$', home),
]

def make_application():
    from django.core.wsgi import get_wsgi_application

    os.environ.setdefault('DJANGO_SETTINGS_MODULE',
                          'wbtframeworks.django.settings')

    return get_wsgi_application()
```

Once all of them are available, we can see that our test is able to run against all four of them without any difference. It can run against **TurboGears2**:

```
$ pytest --framework=tg2
======================= test session starts =======================
platform linux -- Python 3.8.6, pytest-6.0.2, py-1.9.0, pluggy-0.13.1
collected 1 item

tests/test_wsgiapp.py .                                      [100%]
```

```
======================= 1 passed in 0.13s =======================
```

And it can be run against **Django** without any changes:

```
$ pytest --framework=django
======================= test session starts =======================
platform linux -- Python 3.8.6, pytest-6.0.2, py-1.9.0, pluggy-0.13.1
collected 1 item

tests/test_wsgiapp.py .                                        [100%]

======================= 1 passed in 0.13s =======================
```

Now that we are sure that our test suite can run against all four frameworks, we will extend it with the other tests we had for our httpbin.org clone:

```python
import webtest


class TestWSGIApp:
    def test_home(self, wsgiapp):
        client = webtest.TestApp(wsgiapp)
        response = client.get("http://httpbin.org/").text

        assert 'Hello World' in response

    def test_GET(self, wsgiapp):
        client = webtest.TestApp(wsgiapp)
        response = client.get("http://httpbin.org/get").text

        assert '"Host": "httpbin.org"' in response
        assert '"args": {}' in response

    def test_GET_params(self, wsgiapp):
        client = webtest.TestApp(wsgiapp)

        response = client.get(url="http://httpbin.org/get?alpha=1").json

        assert response["headers"]["Host"] == "httpbin.org"
        assert response["args"] == {"alpha": "1"}

    def test_POST(self, wsgiapp):
        client = webtest.TestApp(wsgiapp)

        response = client.post(url="http://httpbin.org/get?alpha=1",
                               params={"beta": "2"}).json

        assert response["headers"]["Host"] == "httpbin.org"
```

```
        assert response["args"] == {"alpha": "1"}
        assert response["form"] == {"beta": "2"}

    def test_DELETE(self, wsgiapp):
        client = webtest.TestApp(wsgiapp)

        response = client.delete(url="http://httpbin.org/anything/27").text
        assert '"method": "DELETE"' in response
        assert '"url": "http://httpbin.org/anything/27"' in response
```

Running the tests now against any framework will complain that those URLs lead to a 404 error, as we haven't yet implemented them. For example, running the tests for **Pyramid** would lead only to the `test_home` one succeeding and the others failing:

```
$ pytest --framework=pyramid
======================= test session starts =======================
platform linux -- Python 3.8.6, pytest-6.0.2, py-1.9.0, pluggy-0.13.1
collected 1 item

tests/test_wsgiapp.py .FFFF                                  [100%]

===================== short test summary info =====================
FAILED test_GET - webtest.app.AppError: 404 Not Found (not 200 OK o...
FAILED test_GET_params - webtest.app.AppError: 404 Not Found (not 2...
FAILED test_POST - webtest.app.AppError: 404 Not Found (not 200 OK ...
FAILED test_DELETE - webtest.app.AppError: 404 Not Found (not 200 O...
================== 4 failed, 1 passed in 0.37s ==================
```

Now that our test suite can run for all four implementations of our application, we only have to proceed with the actual implementation. Given that it doesn't add much value having the same web application implemented in four different frameworks (outside of being a good exercise to learn those frameworks), we are going to provide only the implementation using Django and will leave to the readers the work of implementing it on the other three frameworks if they wish.

Thus we are going to open our `src/wbtframeworks/django/__init__.py` file and edit it to add the remaining routes with the pieces that are lacking:

```python
import sys, json
from django.conf.urls import re_path
from django.conf import settings
from django.http import HttpResponse

settings.configure(
    DEBUG=True,
    ROOT_URLCONF=sys.modules[__name__],
    ALLOWED_HOSTS=["httpbin.org"]
```

```
)

def home(request):
    return HttpResponse('Hello World')

def get(request):
    if request.META.get("SERVER_PORT") == "80":
        host_no_default_port = request.META["HTTP_HOST"].replace(":80", "")
        request.META["HTTP_HOST"] = host_no_default_port
    host = request.META["HTTP_HOST"]

    response = HttpResponse(json.dumps({
        "method": request.META["REQUEST_METHOD"],
        "headers": {"Host": host},
        "args": {
            p: v for (p, v) in request.GET.items()
        },
        "form": {
            p: v for (p, v) in request.POST.items()
        },
        "url": request.build_absolute_uri()
    }, sort_keys=True))
    response['Content-Type'] = 'application/json'
    return response

urlpatterns = [
    re_path(r'^get$', get),
    re_path(r"^anything.*$", get),
    re_path(r'^$', home),
]

def make_application():
    import os
    from django.core.wsgi import get_wsgi_application

    os.environ.setdefault('DJANGO_SETTINGS_MODULE',
                          'wbtframeworks.django.settings')

    return get_wsgi_application()
```

Running our tests now would confirm that, at least for Django, they are able to pass and succeed:

```
$ pytest --framework=django
======================= test session starts =======================
platform linux -- Python 3.8.6, pytest-6.0.2, py-1.9.0, pluggy-0.13.1
collected 5 items
```

```
tests/test_wsgiapp.py .....                                    [100%]

======================= 5 passed in 0.27s =======================
```

This is a very naïve and basic implementation for the sole purpose of showing that our tests are able to pass once the application is provided, but it proves that on Django, we are perfectly able to use WebTest like we would for any other WSGI framework.

But **WebTest** is not the only way we can test Django applications. **Django** also provides its own testing client, so let's see how we would test the same application using Django's test client instead of WebTest.

Writing Django tests with Django's test client

While on Python the most widespread testing toolkit is `pytest`, some web frameworks provide their own solutions for managing test suites. Django is one such example, even though it's possible (as we have seen in the previous section), most people tend to run their tests with the Django test client, which provides the same capabilities as WebTest but is a solution built explicitly for Django.

In this section, we are going to see how we can create a Django project and then run its tests using the standard Django testing infrastructure as well as a pytest-based one:

1. The first step will be to create a new Django project, which we are going to call `djapp`:

   ```
   $ django-admin startproject djapp
   ```

   This will create a `djapp` directory where we can manage our Django project. In the project directory, we will find a `manage.py` file, which allows us to run various management operations for our project, from setting up the database to starting the web application itself and running tests the Django way.

2. The next step is to actually put an application inside our project. As our application will be the `httpbin` one we already wrote, we will just call the application `httpbin`. To create a new application inside a project, we can use the `manage.py startapp` command:

   ```
   $ python manage.py startapp httpbin
   ```

3. Now that the `httpbin` application is available, we have to copy the content of the `wbtframeworks/django/__init__.py` file we just wrote in the previous section. The first things we have to copy are the two `home` and `get` views, which have to be copied inside the `djapp/httpbin/views.py` file:

```python
import json

from django.http import HttpResponse


def home(request):
    return HttpResponse('Hello World')

def get(request):
    if request.META.get("SERVER_PORT") == "80":
        http_host = request.META.get("HTTP_HOST", "httpbin.org")
        host_no_default_port = http_host.replace(":80", "")
        request.META["HTTP_HOST"] = host_no_default_port
    host = request.META["HTTP_HOST"]

    response = HttpResponse(json.dumps({
        "method": request.META["REQUEST_METHOD"],
        "headers": {"Host": host},
        "args": {
            p: v for (p, v) in request.GET.items()
        },
        "form": {
            p: v for (p, v) in request.POST.items()
        },
        "url": request.build_absolute_uri()
    }, sort_keys=True))
    response['Content-Type'] = 'application/json'
    return response
```

4. Then, once the views are available, we must actually expose them; that is, make them accessible through some kind of URL. To do so, we have to add the three URL paths to the `djapp/httpbin/urls.py` file:

```python
from django.urls import re_path

from . import views

urlpatterns = [
    re_path(r'^get$', views.get),
    re_path(r"^anything.*$", views.get),
    re_path(r'^$', views.home)
]
```

Our application is now fully functional. But if we try to start it now it won't work. That's because we haven't yet attached the application to the project. So the djapp project doesn't yet know that it has to serve the httpbin application.

5. To do this, we can open the djapp/djapp/urls.py file and make sure that all the URLs from the httpbin project are correctly included in it:

```
from django.contrib import admin
from django.urls import path, include


urlpatterns = [
    path('admin/', admin.site.urls),
    path("", include("httpbin.urls"))
]
```

6. The last step is to make sure that our website is accessible on all the hosts that we plan to use, so we should set the ALLOWED_HOSTS variable in djapp/djapp/settings.py:

```
ALLOWED_HOSTS = ["httpbin.org", "127.0.0.1"]
```

If we did everything correctly, running manage.py runserver should now run our website and make it visible on http://127.0.0.1:8000/:

```
$ python manage.py runserver
...
Django version 3.1.4, using settings 'djapp.settings'
Starting development server at http://127.0.0.1:8000/
Quit the server with CONTROL-C.
```

Pointing our web browser to http://127.0.0.1:8000/ should greet us with a **Hello World** message as specified by the httpbin.views.home function:

Figure 11.2 – Hello World response from our Django application

Now that we confirmed the application is being correctly served, we have to make sure we are able to run the tests against it.

Testing Django projects with pytest

The first thing we are going to do is to take our test suite as-is, based on WebTest and pytest, and make it work against the new Django project we just wrote. This mostly guarantees that the behavior we have is the same exact behavior we previously had, as the tests are the same tests we had previously. Also shows how we can use pytest and WebTest even with a full-fledged Django project.

To do so, we are going to create a `pytest-tests` directory inside the `djapp` project. Here we are going to place the `djapp/pytest-tests/test_djapp.py` module, which is mostly a copy of the test module we had in the previous section. The only difference will be where the `wsgiapp` object comes from:

```
import sys
import webtest

sys.path.append(".")
from djapp.wsgi import application as wsgiapp


class TestWSGIApp:
    def test_home(self):
        client = webtest.TestApp(wsgiapp)
        response = client.get("http://httpbin.org/").text

        assert 'Hello World' in response

    def test_GET(self):
        client = webtest.TestApp(wsgiapp)
        response = client.get("http://httpbin.org/get").text

        assert '"Host": "httpbin.org"' in response
        assert '"args": {}' in response

    def test_GET_params(self):
        client = webtest.TestApp(wsgiapp)

        response = client.get(url="http://httpbin.org/get?alpha=1").json

        assert response["headers"]["Host"] == "httpbin.org"
        assert response["args"] == {"alpha": "1"}

    def test_POST(self):
        client = webtest.TestApp(wsgiapp)

        response = client.post(url="http://httpbin.org/get?alpha=1",
```

```
                                    params={"beta": "2"}).json

        assert response["headers"]["Host"] == "httpbin.org"
        assert response["args"] == {"alpha": "1"}
        assert response["form"] == {"beta": "2"}

    def test_DELETE(self):
        client = webtest.TestApp(wsgiapp)

        response = client.delete(url="http://httpbin.org/anything/27").text
        assert '"method": "DELETE"' in response
        assert '"url": "http://httpbin.org/anything/27"' in response
```

The two main changes compared to the prior test module are that we removed all the wsgiapp arguments from the test functions, as the wsgiapp object won't come anymore from a fixture injecting the dependency, and that we imported it at the top of the file from the djapp.wsgi module. Different to most web frameworks, in Django the projects are not Python distributions, and thus can't be installed with pip. This means that we can't directly import the project from anywhere and refer to its content.

To surpass this limitation we are going to use sys.path.append(".") to make the current path available to Python. This allows us to import the djapp package inside the djapp project as if it were a normal Python installed package, thus making accessible the djapp.wsgi module. Inside that module, Django makes the WSGI application available as the application object.

To confirm things worked as expected, we are going to run pytest and point it to the pytest-tests directory. This should run the same exact tests we had before, just against the new Django project:

```
$ pytest pytest-tests -v
========================= test session starts =========================
platform linux -- Python 3.8.6, pytest-6.0.2, py-1.9.0, pluggy-0.13.1 --
collected 5 items

pytest-tests/test_djapp.py::TestWSGIApp::test_home PASSED [ 20%]
pytest-tests/test_djapp.py::TestWSGIApp::test_GET PASSED [ 40%]
pytest-tests/test_djapp.py::TestWSGIApp::test_GET_params PASSED [ 60%]
pytest-tests/test_djapp.py::TestWSGIApp::test_POST FAILED [ 80%]
pytest-tests/test_djapp.py::TestWSGIApp::test_DELETE FAILED [100%]

============================== FAILURES ==============================
...
---------------------- Captured stderr call ----------------------
Forbidden (CSRF cookie not set.): /anything/27
---------------------- Captured log call ----------------------
```

```
WARNING django.security.csrf:log.py:224 Forbidden (CSRF cookie not set.):
/anything/27
====================== short test summary info =======================
FAILED pytest-tests/test_djapp.py::TestWSGIApp::test_POST - webtest...
FAILED pytest-tests/test_djapp.py::TestWSGIApp::test_DELETE - webte...
==================== 2 failed, 3 passed in 0.40s =====================
```

Surprisingly, there were two tests that failed compared to before: `test_POST` and `test_DELETE`.

Both of them failed with a **CSRF cookie not set** error. This is because Django sets up support for CSRF attack protection by default in all new projects. The protection works by using a token provided automatically by forms when they get submitted to other endpoints. The problem is that in our project, we don't have any forms at all, so the `DELETE` and `POST` requests are not submitting any tokens, thus failing the protection check.

For our kind of application, this kind of protection doesn't make much sense, as we aren't going to have any forms present. Thus we can edit the `djapp/djapp/settings.py` file and remove the `django.middleware.csrf.CsrfViewMiddleware` line from the `MIDDLEWARES` variable:

```
MIDDLEWARE = [
    'django.middleware.security.SecurityMiddleware',
    'django.contrib.sessions.middleware.SessionMiddleware',
    'django.middleware.common.CommonMiddleware',
#   'django.middleware.csrf.CsrfViewMiddleware',
    'django.contrib.auth.middleware.AuthenticationMiddleware',
    'django.contrib.messages.middleware.MessageMiddleware',
    'django.middleware.clickjacking.XFrameOptionsMiddleware',
]
```

Now that we have disabled the CSRF protection, rerunning the tests should succeed as expected:

```
$ pytest pytest-tests -v
======================= test session starts =========================
collected 5 items

pytest-tests/test_djapp.py::TestWSGIApp::test_home PASSED [ 20%]
pytest-tests/test_djapp.py::TestWSGIApp::test_GET PASSED [ 40%]
pytest-tests/test_djapp.py::TestWSGIApp::test_GET_params PASSED [ 60%]
pytest-tests/test_djapp.py::TestWSGIApp::test_POST PASSED [ 80%]
pytest-tests/test_djapp.py::TestWSGIApp::test_DELETE PASSED [100%]


======================= 5 passed in 0.21s ===========================
```

Testing Django projects with Django's test client

We have seen that the application behaves like we expected as the same tests we wrote for WebTest worked correctly. But running tests using pytest and WebTest is a non-standard way to test Django projects. Most people would expect to be able to test a Django project simply using the `manage.py test` command. But for now, this command runs no tests at all:

```
$ python manage.py test
System check identified no issues (0 silenced).

----------------------------------------------------------------------
Ran 0 tests in 0.000s

OK
```

This is because for `manage.py test` itself, we have not yet written any test. `manage.py test` is mostly based on the `unittest` framework we saw at the beginning of the book, and thus is not compatible with `pytest`. Also, the tests here are meant to be written slightly differently without using `WebTest`.

To migrate our tests to the Django way, we have to create a `djapp/httpbin/tests.py` file in which we will put all our tests. For now, in this file, we are going to provide a single test for the index page of the website, just to make sure that the test suite is able to find our test and that the web application is correctly starting up:

```python
from django.test import TestCase


class HttpbinTests(TestCase):
    def test_home(self):
        response = self.client.get("/")
        self.assertContains(response, "Hello World")
```

Django tests will usually inherit from `django.test.TestCase`, which serves two different purposes:

- First, to make sure that the methods inside the subclass are correctly identified as tests, and thus run when we start the test suite.
- The second purpose is to provide the `self.client` object, which helps to perform requests to the web application much like WebTest did.

In this case, the primary difference is that the web application is not explicitly provided to the client, but is detected based on the project where we are running the tests.

Now that we have a test in place, running the `manage.py test` command again should finally find and run the test:

```
$ python manage.py test
Creating test database for alias 'default'...
System check identified no issues (0 silenced).
.
----------------------------------------------------------------------
Ran 1 test in 0.004s

OK
Destroying test database for alias 'default'...
```

The next steps will be to also port the `test_GET`, `test_GET_params`, `test_POST`, and `test_DELETE` tests to the Django standard so that our full test suite is available when running `manage.py test`.

The main differences when working with Django's test client is that for the responses, it's going to provide a Django `HttpResponse` object, thus the content of the response will be available in binary form in the `HttpResponse.content` attribute and we will have to decode it ourselves, while WebTest provided the `.text` and `.json` properties, which handled much of that for us. Apart from these minor differences, the tests mostly look the same as before:

```python
import json

from django.test import TestCase


class HttpbinTests(TestCase):
    def test_home(self):
        response = self.client.get("/")
        self.assertContains(response, "Hello World")

    def test_GET(self):
        response = self.client.get("/get").content.decode("utf-8")

        assert '"Host": "httpbin.org"' in response
        assert '"args": {}' in response

    def test_GET_params(self):
        response = json.loads(self.client.get("/get?alpha=1").content)

        assert response["headers"]["Host"] == "httpbin.org"
        assert response["args"] == {"alpha": "1"}
```

```
def test_POST(self):
    response = json.loads(self.client.post(
        "/get?alpha=1", {"beta": "2"}
    ).content)

    assert response["headers"]["Host"] == "httpbin.org"
    assert response["args"] == {"alpha": "1"}
    assert response["form"] == {"beta": "2"}

def test_DELETE(self):
    response = self.client.delete(
        "/anything/27"
    ).content.decode("utf-8")
    assert '"method": "DELETE"' in response
    assert '"url": "http://httpbin.org/anything/27"' in response
```

Now that we have in place the same tests we had before, but now in the new Django test client format, we can verify that all five of them pass as expected by rerunning the `manage.py test` command:

```
$ python manage.py test
Creating test database for alias 'default'...
System check identified no issues (0 silenced).
.....
-------------------------------------------------------------------
Ran 5 tests in 0.010s

OK
Destroying test database for alias 'default'...
```

This confirms that our tests for the application succeed also in the Django test client version.

Which format to use (WebTest or Django's test client) can be considered mostly a matter of preference for skilled Django users, but for most developers out there, using Django's test client will probably lead to finding more answers to your questions and doubts, as it's the documented and suggested way that Django developers expect will be used when writing Django applications.

For people interested in using pytest with Django, a `pytest-django` package also exists that tries to fill the gap while hiding most of the machinery necessary to make Django tests run with pytest.

Summary

In this chapter, we saw how we can test HTTP-based applications and how we can verify the behavior of HTTP clients, HTTP servers, and even the two of them together. This is all thanks to the WSGI protocol that powers the Python web ecosystem. We have also seen how testing works in the Django world when Django's test client is used, thus we are fairly capable of writing effective test suites for whatever web framework we are going to use.

Our testing isn't fully complete by the way. We are verifying the endpoints, checking that the web pages contain the responses we expect, but we have no way to check that, once those responses are read by a web browser, they actually behave as we expected. Even worse, if there is JavaScript involved, we don't have any way to verify that the JavaScript in those web pages is actually doing what we want.

So in the next chapter, we are going to see how we can test our web applications with a real browser while also verifying the JavaScript that our web pages contain, thus completing the list of skills we might need to develop a fully tested web application.

12
End-to-End Testing with the Robot Framework

In the previous chapter, we saw how to test web applications and, in general, applications that rely on the HTTP protocol, both client and server side, but we were unable to test how they perform in a real browser. With their complex layouts, the fact that CSS and JavaScript are heavily involved in testing your application with WebTest or a similar solution might not be sufficient to guarantee users that they are actually able to work with it. What if a button is created by JavaScript or it's disabled by CSS? Those conditions are hard to test using WebTest and we might easily end up with a test that clicks that button even though the button wasn't actually usable by users.

To guarantee that our applications behave properly, it is a good idea to have a few tests that verify at least the more important areas of the application using a real browser. As those kinds of tests tend to be very slow and fragile, you still want to have the majority of your tests written using solutions such as WebTest or even unit tests, which don't involve the whole application life cycle, but having the most important parts of the web application verified using real browsers will guarantee that at least the critical path of your web application works on all major browsers.

The Robot framework is one of the most solid solutions for writing the end-to-end tests that drive web browsers and mobile applications in the Python world. It was originally developed by Nokia and evolved under the open source community, and is a long-standing and solid solution with tons of documentation and plugins. It is therefore battle tested and ready for your daily projects.

In this chapter, we will cover the following topics:

- Introducing the Robot framework
- Testing with web browsers
- Extending the Robot framework

Technical requirements

We need a working Python interpreter with the Robot Framework installed. To run tests with real browsers, we are also going to use the `robotframework-seleniumlibrary` and the `webdrivermanager` utilities. To record videos of our tests, we are going to need the `robotframework-screencaplibrary` library. `robotframework`, `robotframework-seleniumlibrary`, `robotframework-screencaplibrary`, and `webdrivermanager` can be installed with `pip`, in the same way as all other Python dependencies:

```
$ pip install robotframework robotframework-seleniumlibrary
webdrivermanager robotframework-screencaplibrary
```

The examples have been written on Python 3.7, robotframework 3.2.2, robotframework-seleniumlibrary 4.5.0, robotframework-screencaplibrary 1.5.0, and webdrivermanager 0.9.0, but should work on most modern Python versions.

You can find the code present in this chapter on GitHub at `https://github.com/PacktPublishing/Crafting-Test-Driven-Software-with-Python/tree/main/Chapter12`.

Introducing the Robot Framework

The Robot Framework is an automation framework mostly used to create acceptance tests in the **Acceptance Test Driven Development (ATDD)** and **Behavior Driven Development (BDD)** styles. Tests are written in a custom, natural English-like language that can be easily extended in Python, so Robot can, in theory, be used to write any kind of acceptance tests in a format that can be shared with other stakeholders, pretty much like what we have seen we can do with `pytest-bdd` in previous chapters.

The primary difference is that Robot is not based on PyTest, it is a replacement for PyTest, and is widely used to create end-to-end tests for mobile and web applications. For mobile applications, the Appium library allows us to write Robot Framework tests that control mobile applications on a real device, while the Selenium library provides a complete integration with web browsers, which means that the Robot Framework allows us to write tests that drive a real web browser and verify the results.

Robot Framework tests are written inside `.robot` files, which are then divided into multiple sections by the section headers. The most frequently used section headers are the following:

- `*** Settings ***`: This contains options to configure Robot itself.
- `*** Variables ***`: This contains variables to reuse across multiple tests.

- `*** Test Cases ***`: This contains our tests.
- `*** Keywords ***`: This contains our own custom commands.

So, the minimum content of a `.robot` file is usually a `Test Cases` section with a test inside. Each test is then a collection of commands for the Robot Framework that are provided by keywords made available by libraries for the Robot Framework itself.

The only library automatically available by default in the Robot Framework is the `Builtin` one (Builtin library reference: `https://robotframework.org/robotframework/latest/libraries/BuiltIn.html`), which provides some generally helpful commands such as `Should Contain` to check the content of a variable, or `Expression` to run any Python expression and assign the result to a variable.

Further libraries can be imported explicitly with the `Library` command in the `Settings` section. Without involving explicit libraries that add more commands, Robot itself can't do much.

For example, if we want to create a very basic test where we save the `"Hello World"` string into a file and verify its content, we would have to involve the `OperatingSystem` library (OperatingSystem library reference: `https://robotframework.org/robotframework/latest/libraries/OperatingSystem.html`), which makes available commands to interact with files, directories, and the system shell.

To create such a test, we would make a `hellotest.robot` file, where we can declare the instruction for the Robot Framework. At the beginning of the file, we would declare a `Settings` section, where we use the `Library` command to make the `OperatingSystem` library available:

```
*** Settings ***
Library       OperatingSystem
```

In Robot, multiple spaces perform separate commands from their arguments.

Through the `OperatingSystem` library, we will get the `Run` and `Get File` commands, which we need to write our actual test.

Subsequently, we will declare the `Test Cases` section, where we can put all our tests. In this case, we are going to place only one test, entitled `Hello World`.

The test itself will create a new file with the "Hello World" string inside, and will then read it back and check that the content contains the string "Hello":

```
*** Test Cases ***
Hello World
    Run    echo "Hello World" > hello.txt
    ${filecontent} =    Get File    hello.txt
    Should Contain    ${filecontent}    Hello
```

The first line of our test uses the Run keyword to invoke the echo command in a shell (if you are on a *nix system, such as Linux or macOS X), and the echo command is invoked with the Hello World argument and redirection is effected to the hello.txt file so that the output of the command actually goes into that file.

Once that file is created, on the second line we use the Get File keyword to read the content of the hello.txt file and assign what we read to the ${filecontent} variable.

Finally, we check through the Should Contain keyword that the variable contains the string Hello.

Once we have saved all this as hellotest.robot, we should be able to run it by invoking the robot command and see that our test is executed and passes:

```
$ robot hellotest.robot
==========================================================
Hellotest
==========================================================
Hello World                                       | PASS |
----------------------------------------------------------
Hellotest                                         | PASS |
1 critical test, 1 passed, 0 failed
1 test total, 1 passed, 0 failed
==========================================================
```

If we wanted to see what happens when our tests fail, we could change the Should Contain line to a different string, for example, Should Contain ${filecontent} Bye and see what happens when we rerun our test:

```
$ robot hellotest.robot
==========================================================
Hellotest
==========================================================
Hello World                                       | FAIL |
'Hello World' does not contain 'Bye'
----------------------------------------------------------
Hellotest                                         | FAIL |
```

```
1 critical test, 0 passed, 1 failed
1 test total, 0 passed, 1 failed
============================================================
```

Details about what precisely went wrong are then made available in the `log.html` file, where each command that Robot performed is recorded with debugging information. Opening such a file in a browser will indicate explicitly that the command that failed is `Builtin.Should Contain` and that it failed with the `'Hello World' does not contain 'Bye'` error:

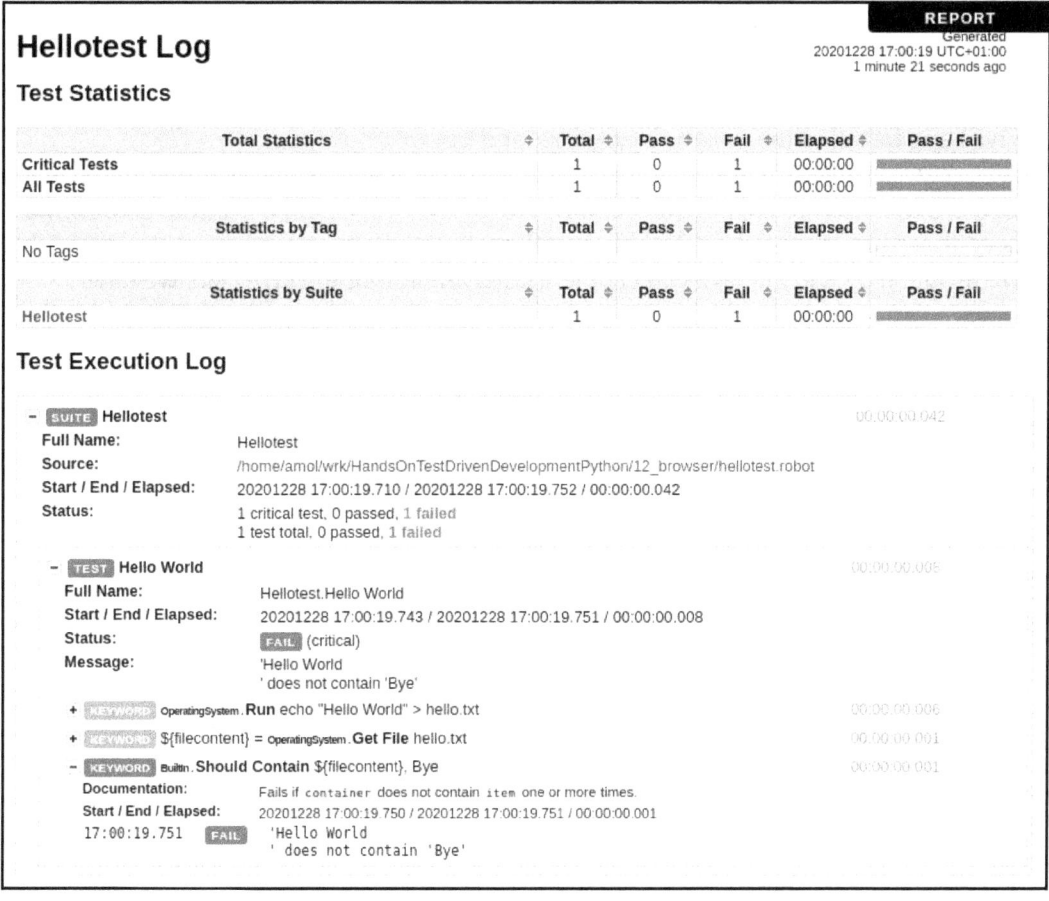

Figure 12.1 – Detailed log of our test execution from log.html

Now that we know how the Robot Framework works, we can move on to the next steps and see how we can use it to test web applications with a real browser.

Testing with web browsers

We have seen how, using libraries, we can extend Robot with additional commands that allow us to write most different kinds of tests. One of the most frequent use cases for Robot is actually web development as it has a very convenient `SeleniumLibrary` library that provides many commands to control a real web browser and perform tests that can involve JavaScript (Selenium library reference: `https://robotframework.org/SeleniumLibrary/SeleniumLibrary.html`).

Once we have installed the `robotframework` and `robotframework-seleniumlibrary` Python distributions, in order to be able to write tests that involve a real browser, we will need to enable the web drivers for the browsers we want to use. So, we will need those browsers to be available and then, through the `webdrivermanager` utility that we installed previously, we can enable the drivers for all the browsers we have available:

```
$ webdrivermanager firefox chrome
Downloading WebDriver for browser: "firefox"
2588kb [00:01, 1978.35kb/s]
Driver binary downloaded to:
"./venv/WebDriverManager/gecko/v0.28.0/geckodriver-v0.28.0-
linux64/geckodriver"
Symlink created: ./venv/bin/geckodriver

Downloading WebDriver for browser: "chrome"
5979kb [00:01, 3615.18kb/s]
Driver binary downloaded to:
"./venv/WebDriverManager/chrome/87.0.4280.88/chromedriver_linux64/chromedri
ver"
Symlink created: ./venv/bin/chromedriver
```

 Notice that the examples take for granted the fact that everything is happening inside a Python virtual environment, so keep in mind that when using a virtual environment, the drivers are only available inside that environment, and if you create a new one you will need to enable the drivers again.

Once we have the drivers available, Robot will be able to control the browsers for which we provided the drivers (in this case, Chrome and Firefox), so we can go back to our editor and create a new test to establish how Robot works.

In this case, we are going to create a test where we search on Google for a famous person and verify that Wikipedia is included in the results returned. To do so, let's create a `searchgoogle.robot` file were we are going to enable the `SeleniumLibrary` library so that browser-related commands become available:

```
*** Settings ***
Library SeleniumLibrary
```

The next step is then to write the test itself to open Google with Chrome, accept the privacy policy, perform the search, and then check that Wikipedia is included in the results:

```
*** Test Cases ***
Search On Google
    Open Browser     http://www.google.com    Chrome
    Wait Until Page Contains Element    cnsw
    Select Frame    //iframe
    Submit Form    //form
    Input Text     name=q    Stephen\ Hawking
    Press Keys     name=q    ENTER
    Page Should Contain    Wikipedia
    Close Window
```

If we run our test, a Chrome window will pop up, perform the search, and then close again, with our test regarded as having passed if everything went right:

```
$ robot searchgoogle.robot
==========================================================
Searchgoogle
==========================================================
Search On Google                              | PASS |
----------------------------------------------------------
Searchgoogle                                  | PASS |
1 critical test, 1 passed, 0 failed
1 test total, 1 passed, 0 failed
==========================================================
```

Our test might look a bit complex, and that's because the Google website requires us to accept a privacy policy before we can start searching. So, the first commands are related to opening Google itself using Chrome and then waiting for the privacy policy (with `id=cnsw` in HTML) to appear:

```
    Open Browser     http://www.google.com    Chrome
    Wait Until Page Contains Element    cnsw
```

Once the browser opens the Google website, we should be greeted by the privacy policy acceptance box:

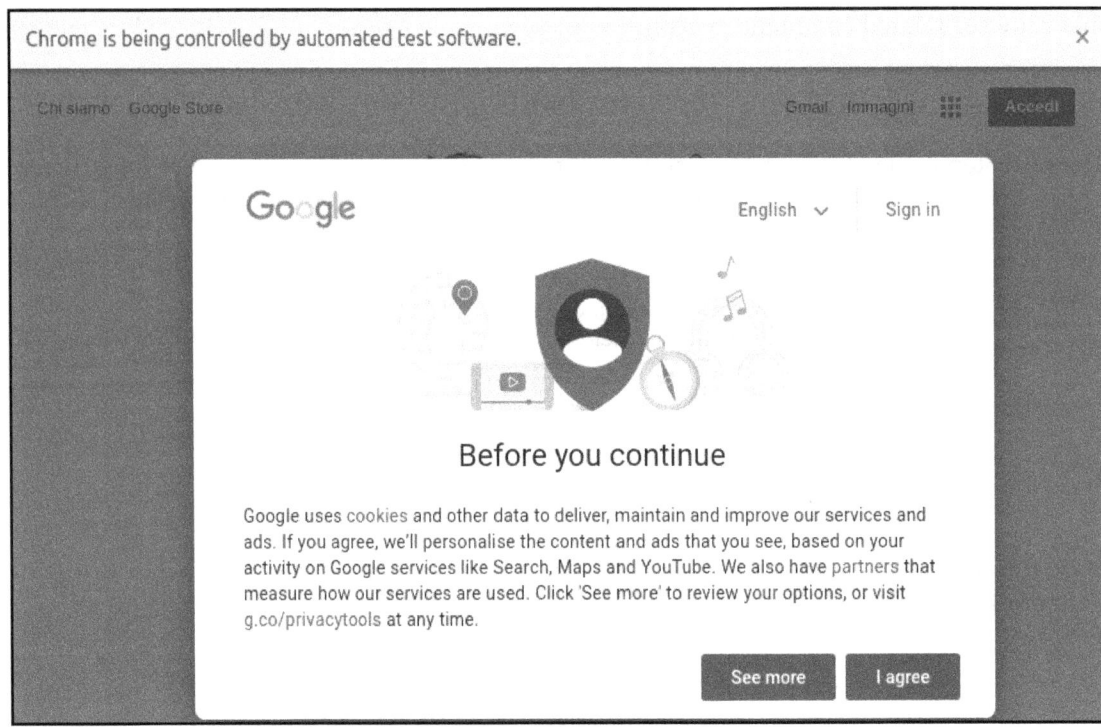

Figure 12.2 – Google website with the policy acceptance request

Figure 12.2 – Google website with the policy acceptance request

In case you don't see the privacy policy when opening the Google website, don't worry. Google decides if to show the privacy policy or not based on the country and browser you are connecting from. If you country doesn't have any privacy policy requirement, Google might not show the policy. In such case you can omit the three "Wait Until Page Contains Element", "Select Frame" and "Submit Form" commands related to managing the privacy policy or just read further until we tackle headless browser later in the chapter and run the examples using Google Chrome browser in headless mode.

Once the privacy policy is visible, we are going to pick the `iframe` within which it gets displayed and submit the first form that exists within it:

```
Select Frame    //iframe
Submit Form     //form
```

Submitting the form will make the privacy policy alert disappear and will finally reveal the search box:

Figure 12.3 – Google website once the privacy policy has been accepted

At this point, we just have to write the name of the person we want to search for in the search box (which has `name=q` in HTML) and submit it by pressing the ENTER key:

```
Input Text    name=q    Stephen\ Hawking
Press Keys    name=q    ENTER
```

Notice that we had to escape the space between the first name and surname of Stephen Hawking, and that's because spaces are used to separate arguments of commands in Robot, so we wanted the name and surname to figure together as a single argument of the `Input Text` command instead of them being treated as separate arguments.

At this point, if everything worked correctly, we should see the search results showing Wikipedia as one of them, if not the first:

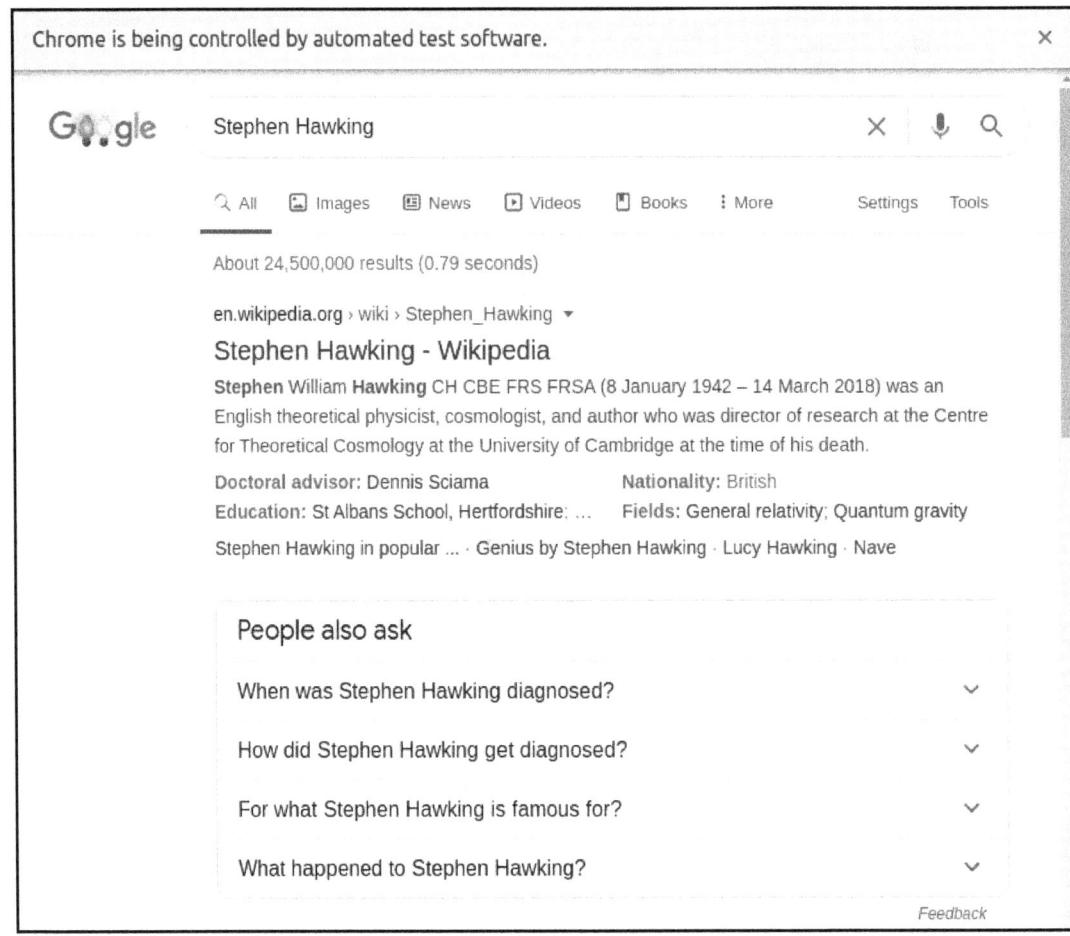

Figure 12.4 – Google search results for "Stephen Hawking"

As we are writing a test, the subsequent line is meant to assert the condition for our test, so it's going to check that Wikipedia is one of the results:

```
Page Should Contain    Wikipedia
```

Once we have verified that everything worked as expected, as we have nothing else to do, we can submit the last command to close the browser window and move forward:

```
Close Window
```

Recording the execution of tests

As we have seen, while tests are running, the browser window is on screen and every action we perform is visible. As we obviously don't want to stare at our tests while they run, it would be convenient to have recordings of them available, so that we can see what happened during those tests in case of a failure.

Luckily for us, the Robot Framework has a `ScreenCapLibrary` library that allows screenshots and video recordings of our tests to be made. Once the `robotframework-screencaplibrary` Python distribution is installed with `pip`, we will be able to use its commands by adding it to our test's `*** Settings ***` section:

```
*** Settings ***
Library    SeleniumLibrary
Library    ScreenCapLibrary
```

To record the execution of a test, we just have to begin it with a `Start Video Recording` command and then end it with a `Stop Video Recording` one:

```
*** Test Cases ***
Search On Google
    Start Video Recording
    Open Browser    http://www.google.com    Chrome
    Wait Until Page Contains Element    cnsw
    Select Frame    //iframe
    Submit Form    //form
    Input Text    name=q    Stephen\ Hawking
    Press Keys    name=q    ENTER
    Stop Video Recording
    Page Should Contain    Wikipedia
    Close Window
```

Screenshots and videos taken with the test are embedded within the `log.html` document, so we can see the result of our recording by looking at the log file:

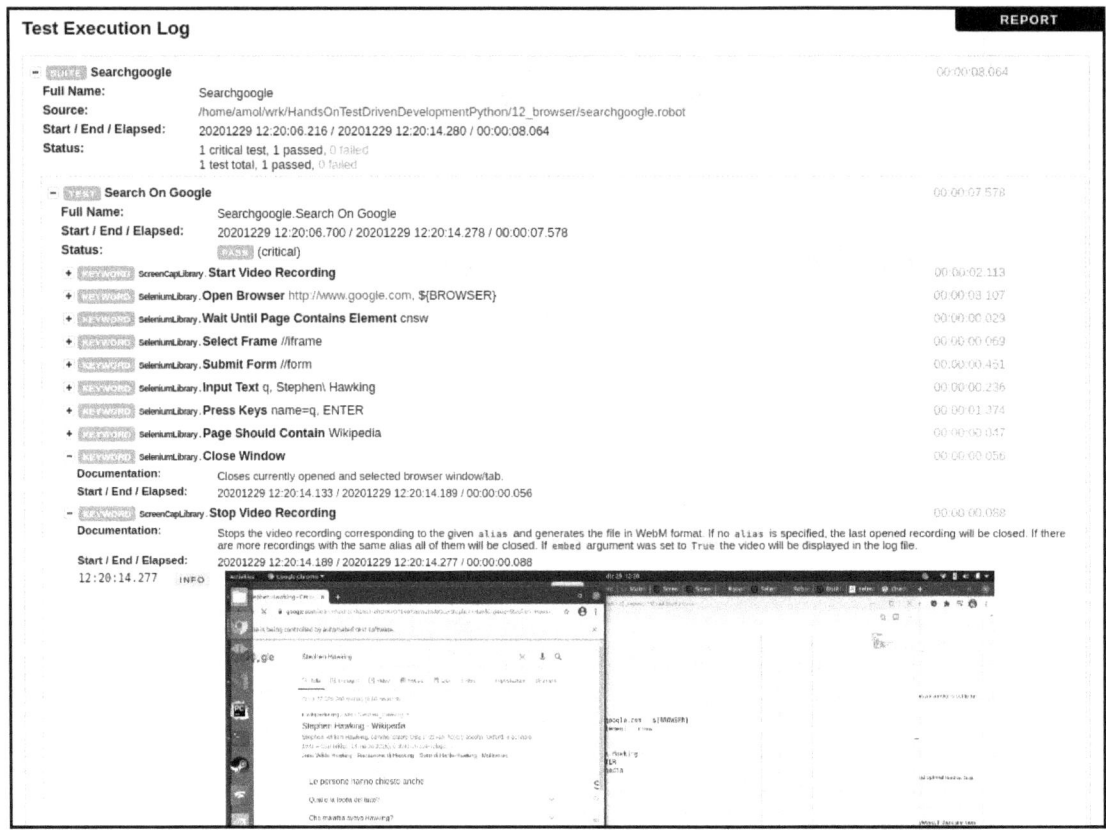

Figure 12.5 – Test execution log with video recording embedded

`ScreenCapLibrary` recordings will be available only if the step that saves them succeeds. Therefore, we need to pay attention when writing our tests to ensure that recordings are saved (which means stopping the recording before any assertion). In our short test, for example, we placed the `Stop Video Recording` command before the `Page Should Contain Wikipedia` one. This ensures that even if Wikipedia is not included in the results, the recording will still be visible:

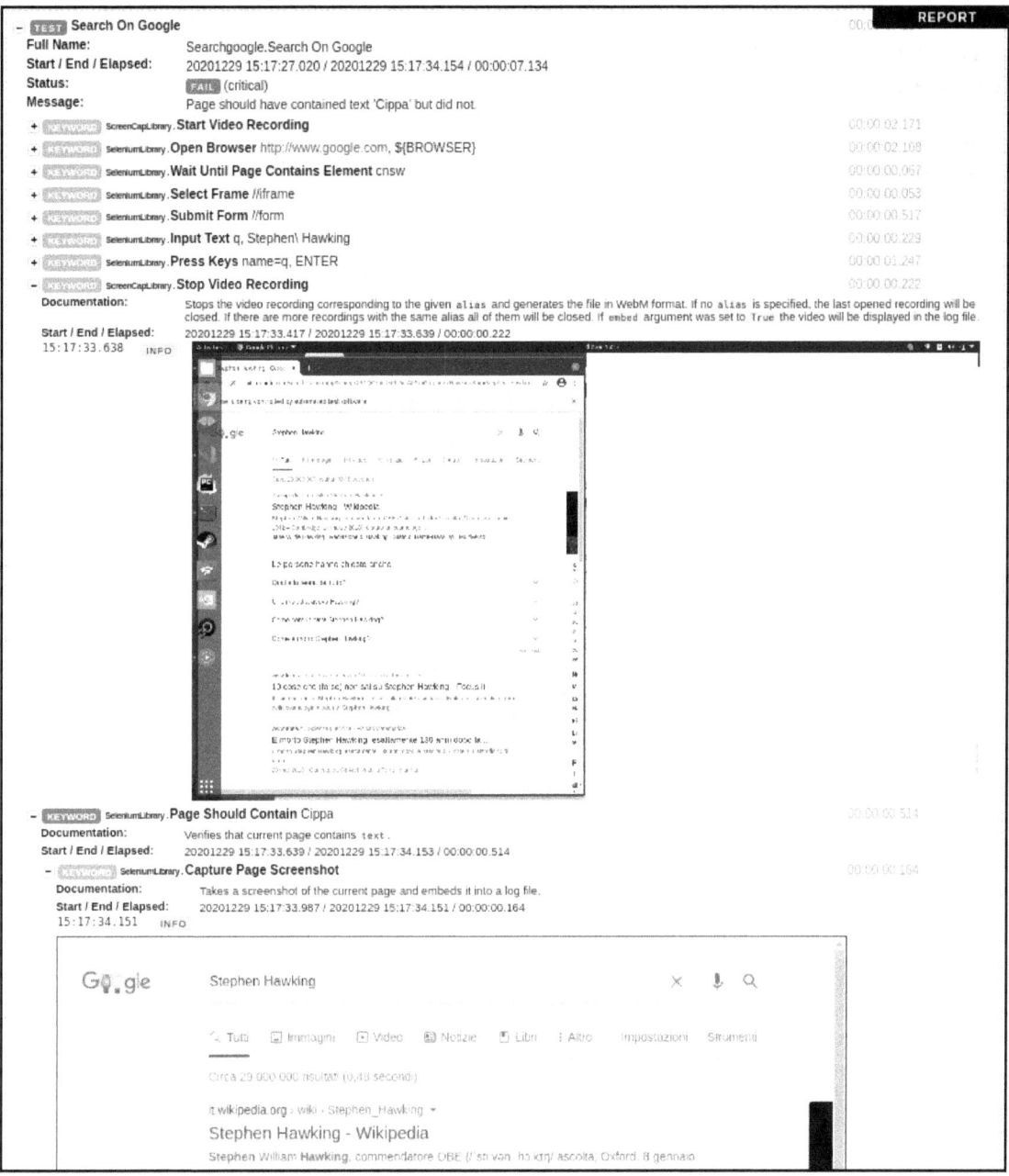

Figure 12.6 – Test execution log with the recording even if the test assertion failed

At the other end, in the event of any failure, the `SeleniumLibrary` library will make a screenshot of the web browser. So, even if our video doesn't get recorded, we will always have available screenshots of the state of the browser at the time the command failed.

A more robust approach for handling recording is to rely on the `Test Setup` and `Test Teardown` phases of Robot so that we can start and stop the recording on every test automatically and even in case of failures. So if, for example, we move our `Start Video Recording` and `Stop Video Recording` commands into those two phases within the `Settings` section, we will have a reliable recording even in the event of failures:

```
*** Settings ***
Library     SeleniumLibrary
Library     ScreenCapLibrary

Test Setup      Start Video Recording
Test Teardown    Stop Video Recording

*** Test Cases ***
Search On Google
    Open Browser     http://www.google.com     Chrome
    Wait Until Page Contains Element     cnsw
    Select Frame     //iframe
    Submit Form     //form
    Input Text     name=q     Stephen\ Hawking
    Press Keys     name=q     ENTER
    Page Should Contain     Wikipedia
    Close Window
```

Now, our recording will be started automatically on all tests and stopped when they end, even if they fail.

It's generally a good idea to make sure that your test suite has a `Suite Teardown` step with a `Close All Browsers` command in the `*** Settings ***` section. This will ensure that all browser processes and windows are properly destroyed when the test suite finishes running. Some browsers tend to leave behind running processes after the tests have run, and so might slow down your system if you run the test suite multiple times.

Testing with headless browsers

Even if it's convenient to be able to see what's going on during tests, during our daily development cycle, we don't want to have browser windows popping up in the middle of our screen and preventing us from doing anything else apart from looking at our tests running.

For this reason, it's frequently convenient to be able to run tests without real browser windows opening. This can be done by using a **headless** browser, in other words, a browser without a UI.

With Chrome, for example, this can be done in the `Open Browser` command by choosing the `headlesschrome` browser instead of `Chrome`. Using `headlesschrome` will prevent browser windows from popping up, but will still retain the majority of the features:

```
*** Test Cases ***
Search On Google
    Open Browser     http://www.google.com     headlesschrome
    Wait Until Page Contains Element    cnsw
    Select Frame    //iframe
    Submit Form     //form
    Input Text      name=q    Stephen\ Hawking
    Press Keys      name=q    ENTER
    Page Should Contain    Wikipedia
    Close Window
```

Unfortunately, while Robot will retain the same behaviors when running with a headless browser, the websites themselves might not. So, for example, in our case, the test will fail because Google won't show up the privacy policy acceptance dialog when running with a headless browser:

```
$ robot searchgoogle.robot
======================================================
Searchgoogle
======================================================
Search On Google                                | FAIL |
Element 'cnsw' did not appear in 5 seconds.
------------------------------------------------------
Searchgoogle                                    | FAIL |
1 critical test, 0 passed, 1 failed
1 test total, 0 passed, 1 failed
======================================================
```

To address this issue, we can make the commands related to the privacy policy conditional and only run them when a normal browser is in use. To do so, the first step is to refactor the selected browser into a variable so that we can more easily change which browser we are going to use:

```
*** Variables ***
${BROWSER}     chrome

*** Test Cases ***
Search On Google
    Open Browser      http://www.google.com      ${BROWSER}
    ...
```

Now that we can easily change which browser we use just by changing the ${BROWSER} variable, we can check whether that variable contains "headlesschrome" to skip the privacy policy part in the case of the Chrome browser in headless mode.

To make an instruction conditional, we can use the Run Keyword If command. Tweaking our test that way will make sure that it succeeds both when using a real browser or a headless one:

```
*** Settings ***
Library     SeleniumLibrary
Library     ScreenCapLibrary
Test Setup Start     Video Recording
Test Teardown Stop     Video Recording

*** Variables ***
${BROWSER}     headlesschrome
${NOTHEADLESS}=     "headlesschrome" not in "${BROWSER}"

*** Test Cases ***
Search On Google
    Open Browser      http://www.google.com      ${BROWSER}
    Run Keyword If     ${NOTHEADLESS}     Wait Until Page Contains Element
        cnsw
    Run Keyword If     ${NOTHEADLESS}     Select Frame     //iframe
    Run Keyword If     ${NOTHEADLESS}     Submit Form     //form
    Input Text     name=q     Stephen\ Hawking
    Press Keys     name=q     ENTER
    Page Should Contain     Wikipedia
    Close Window
```

To avoid repeating the condition over and over, we also refactored the "headlesschrome" not in "${BROWSER}" expression into a variable so that we can just check for that variable.

Now that we have conditional execution of the instructions that caused problems when using a headless browser, we can rerun our test:

```
$ robot searchgoogle.robot
========================================================
Searchgoogle
========================================================
Search On Google                                | PASS |
--------------------------------------------------------
Searchgoogle                                    | PASS |
1 critical test, 1 passed, 0 failed
1 test total, 1 passed, 0 failed
========================================================
```

Now, our tests finally passed using a headless browser and we learned how to use variables and conditional execution in Robot tests.

Testing multiple browsers

Now that we know how to run tests in Chrome, headless or not, it might be reasonable to feel the need to verify that our web application actually works on other browsers, too. So the question might be how we can also verify it on Firefox or Edge.

Luckily for us, we just refactored the browser in use to be a variable, so we can just change that variable and have all our tests run on one browser or the other.

But if we want to make this part of our CI, it's not very convenient to change the tests file in the middle of our CI runs. For this reason, Robot allows the provision of variable values through the command line using the --variable option. For example, to use Firefox, we could pass --variables browser:firefox:

```
$ robot --variable browser:firefox searchgoogle.robot
========================================================
Searchgoogle
========================================================
Search On Google                                | FAIL |
Element with locator 'name=q' not found.
--------------------------------------------------------
Searchgoogle                                    | FAIL |
1 critical test, 0 passed, 1 failed
1 test total, 0 passed, 1 failed
========================================================
```

Surprisingly, when run with Firefox, our test failed. This is not only because websites might behave differently when using different browsers, but also because the browsers themselves might behave differently.

For example, Firefox didn't select back the primary page after we accepted the privacy policy, so it's still trying to act inside the `iframe` that contained the privacy policy. This makes it impossible for the browser to find the `input` with `name=q`, where it's meant to write the query string, and so the test is failing.

To fix this, we can modify our test slightly to `Unselect Frame` after we have finished with it:

```
*** Test Cases ***
Search On Google
    Open Browser      http://www.google.com     ${BROWSER}
    Run Keyword If    ${NOTHEADLESS}    Wait Until Page Contains Element
        cnsw
    Run Keyword If    ${NOTHEADLESS}    Select Frame    //iframe
    Run Keyword If    ${NOTHEADLESS}    Submit Form    //form
    Unselect Frame
    Input Text    name=q    Stephen\ Hawking
    Press Keys    name=q    ENTER
    Page Should Contain    Wikipedia
    Close Window
```

This will make sure that the test is able to accept the privacy policy and go back to the search field in both Chrome and Firefox, thus solving our problem. Now that we are able to perform the search, let's go back to our tests and see what happens when rerunning them:

```
$ robot --variable browser:firefox searchgoogle.robot
==================================================
Searchgoogle
==================================================
Search On Google                          | FAIL |
Page should have contained text 'Wikipedia' but did not.
--------------------------------------------------
Searchgoogle                              | FAIL |
1 critical test, 0 passed, 1 failed
1 test total, 0 passed, 1 failed
==================================================
```

Another apparent failure is that even though the search happened correctly, the browser was unable to find Wikipedia in the results.

In this case, the `log.html` output can immediately help us understand what's going wrong. If we look at it, we will see that the problem is that by the time our test checks for "Wikipedia", the web page has not yet loaded the results themselves. The search box is still visible in the screenshot that the log file contains:

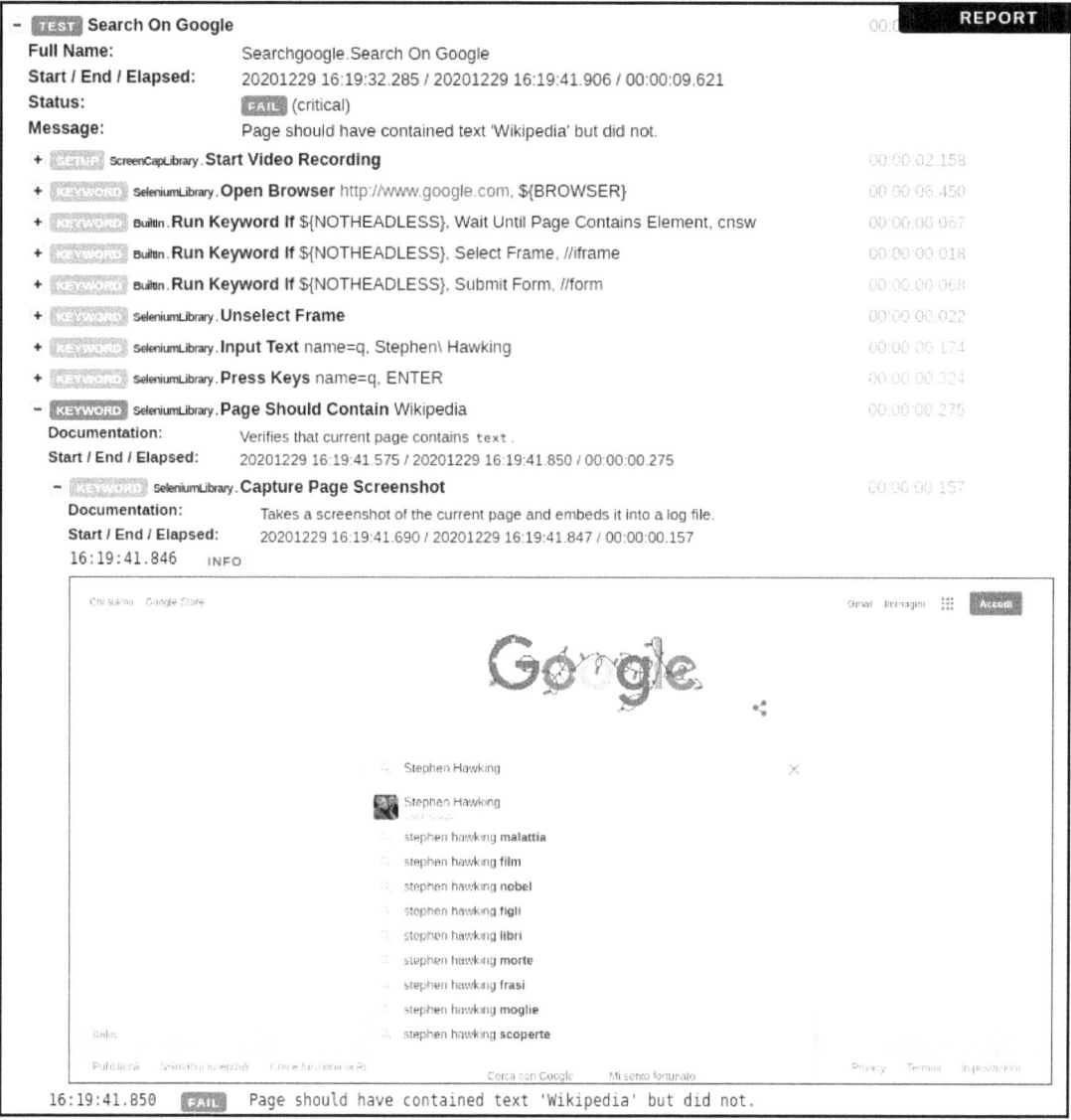

Figure 12.7 – The log for our test that failed because Firefox was slower than the test itself

We can fix this by waiting for the search results to appear before performing the assertion, so let's tweak our search test a bit more to include an explicit wait for the results:

```
*** Test Cases ***
Search On Google
    Open Browser       http://www.google.com      ${BROWSER}
    Run Keyword If     ${NOTHEADLESS}     Wait Until Page Contains Element
        cnsw
    Run Keyword If     ${NOTHEADLESS}     Select Frame     //iframe
    Run Keyword If     ${NOTHEADLESS}     Submit Form      //form
    Unselect Frame
    Input Text     name=q     Stephen\ Hawking
    Press Keys     name=q     SPACE
    Press Keys     name=q     ENTER
    Wait Until Page Contains Element     id=res
    Page Should Contain     Wikipedia
    Close Window
```

This last version of our test is finally able to pass in connection with all the browsers we were concerned with, Firefox and Chrome, with both of them in headless mode too:

```
$ robot --variable browser:headlessfirefox searchgoogle.robot
==============================================================
Searchgoogle
==============================================================
Search On Google                                       | PASS |
--------------------------------------------------------------
Searchgoogle                                           | PASS |
1 critical test, 1 passed, 0 failed
1 test total, 1 passed, 0 failed
==============================================================
```

At this point, we know how to write tests in Robot and how to write them so that we can verify them using multiple different browsers.

Extending the Robot Framework

As we have seen, Robot can be expanded with libraries that can add more keywords. That can be a convenient feature also for us when writing tests. If we have a set of instructions that we are going to repeat frequently in our tests, it would be convenient to factor them into a single keyword that we can reuse. Furthermore, Robot can be expanded with new custom commands that we can implement in Python.

Adding custom keywords

To see how extending Robot with custom keywords works, we are going to create a very simple `customkeywords.robot` test file, where we are going to write a basic script that only greets us:

```
*** Test Cases ***
Use Custom Keywords
    Echo Hello
```

Running the script will fail as we have not yet implemented the `Echo Hello` keyword, so how can we provide it? For this purpose, Robot supports a `*** Keywords ***` section, where we can declare all our custom keywords, so let's declare our keyword there:

```
*** Keywords ***
Echo Hello
    Log Hello!

*** Test Cases ***
Use Custom Keywords
    Echo Hello
```

The `Echo Hello` keyword is just going to invoke the built-in `Log` keyword, passing a hardcoded greeting string, so it's not very helpful, but we could actually list any kind or amount of commands within a custom keyword, so we could make it do whatever we needed.

Now that we have provided a declaration for the `Echo Hello` command, rerunning the tests will succeed:

```
$ robot customkeywords.robot
==================================================
Customkeywords
==================================================
Use Custom Keywords                      | PASS |
--------------------------------------------------
Customkeywords                           | PASS |
1 critical test, 1 passed, 0 failed
1 test total, 1 passed, 0 failed
==================================================
```

The output of our logging is not visible on the shell from which we started the Robot command, but if we open the `log.html` file, we will see that the string was correctly logged in that document.

Extending Robot from Python

Going further, we can expand Robot with new libraries that we can implement in Python. To do so, we have to create a Python package with the name of the library and install it. All the non-internal functions we declare in the installed library will become available in our Robot scripts once we enable the library itself with the usual `Library` command.

So, let's replicate what we just did using Python. The first step is to create the distribution for the library so that it can be installed. Therefore, we are going to create a `hellolibrary` directory where we are going to put our `hellolibrary/setup.py` file:

```
from setuptools import setup

setup(name='robotframework-hellolibrary', packages=['HelloLibrary'])
```

Within this directory, we need to create a `HelloLibrary` package. This will be what gets installed and what gets loaded using the `Library` command in Robot. So let's create a `hellolibrary/HelloLibrary/__init__.py` file so that the nested directory gets recognized as a package by Python.

Inside the `__init__.py` file, we are going to declare the `HelloLibrary` class with a `say_hello` method. The `say_hello` method, as a public method, will be automatically exposed in Robot as the `Say Hello` keyword of the library:

```
class HelloLibrary:

    def say_hello(self):
        print("Hello from Python!")
```

Now that all the pieces are in place, we can install our library so that it becomes available to Robot for installation using `pip`, as we would for any other Python distribution:

```
$ pip install -e hellolibrary/
Obtaining file://hellolibrary
Installing collected packages: robotframework-hellolibrary
  Running setup.py develop for robotframework-hellolibrary
Successfully installed robotframework-hellolibrary
```

Once our library is installed, we can use it as we would for any other Robot library. Adding a `Library HelloLibrary` instruction to our `Settings` section will make the `Say Hello` keyword available for our own use:

```
*** Settings ***
Library HelloLibrary

*** Keywords ***
Echo Hello
    Log Hello!

*** Test Cases ***
Use Custom Keywords
    Echo Hello
    Say Hello
```

We can confirm that everything worked as expected by rerunning Robot. If we didn't make any error and the library was installed correctly, our tests should succeed again:

```
$ robot customkeywords.robot
==================================================
Customkeywords
==================================================
Use Custom Keywords                      | PASS |
--------------------------------------------------
Customkeywords                           | PASS |
1 critical test, 1 passed, 0 failed
1 test total, 1 passed, 0 failed
==================================================
```

Like we did for the `Echo Hello` keyword, we can verify that our `Say Hello` keyword worked properly and logged the `"Hello from Python!"` message by looking at the `log.html` file:

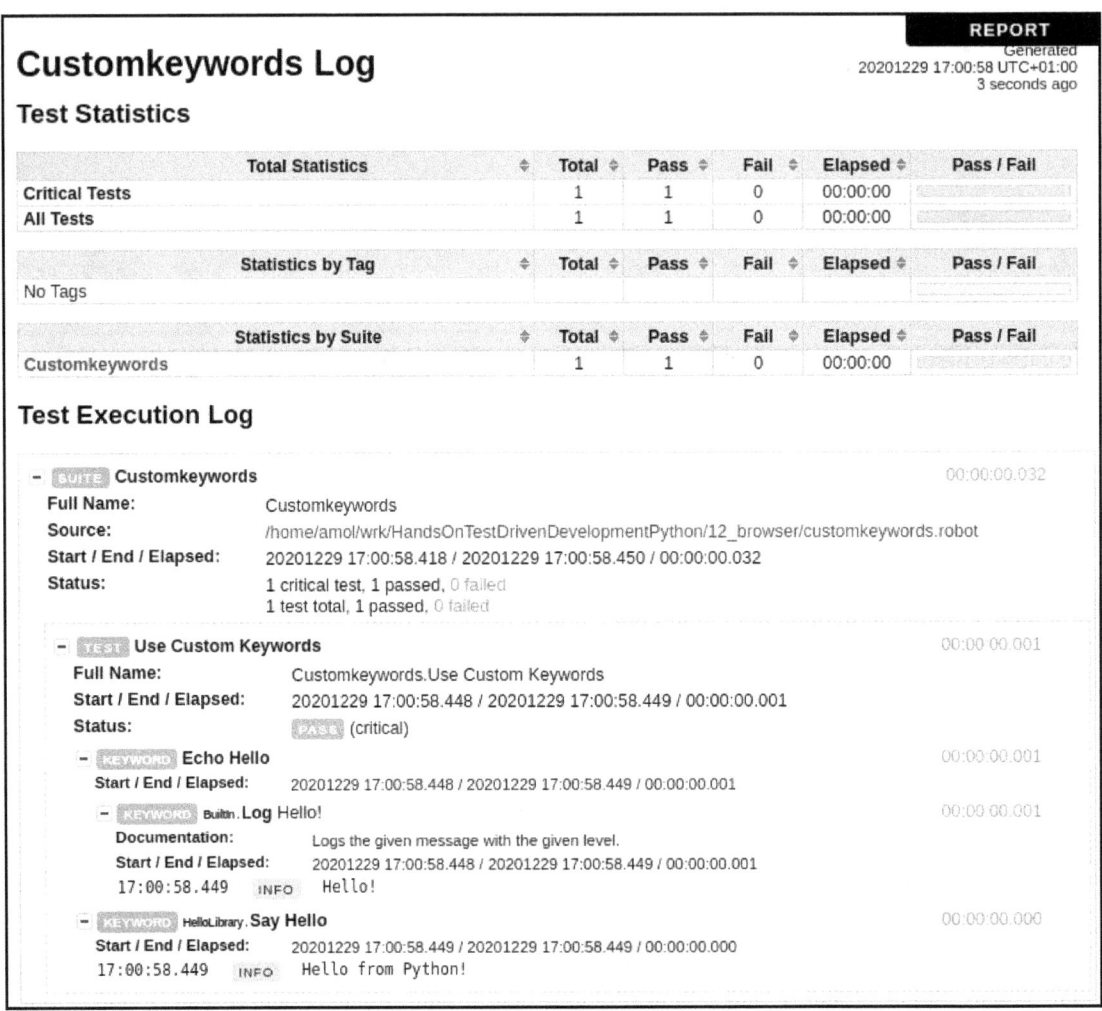

Figure 12.8 – Log of the test using our custom commands

By default, a new library object is created for every test, so a new instance of our `HelloLibrary` class would be made on every test. In case we needed to share a single object across all tests, we could set the `HelloLibrary.ROBOT_LIBRARY_SCOPE = "SUITE"` class attribute, which would signal to Robot to create only once instance and share it across all tests of the same suite. Furthermore, we could set that attribute to `ROBOT_LIBRARY_SCOPE = "GLOBAL"` and make the instance unique for the whole test run. This allows us to share the internal state of our library object across multiple tests in case we need to preserve any information.

Summary

In this chapter, we saw how we can go further and not only test the responses that our web applications provide, but also that those responses work for real once they are handled by a web browser.

Now that we have covered Robot, we have all the tools we need to test our web applications across all stack levels. We know how to use PyTest for building block unit tests, WebTest for functional and integration tests, and Robot for end-to-end tests involving real browsers. So we are now able to write fully tested web applications, paired with the best practices for TDD and ATDD, which we learned in earlier chapters, and we should be able to build a solid development routine that allows us to create robust web applications that are also safe to evolve and refactor over time.

Other Books You May Enjoy

If you enjoyed this book, you may be interested in these other books by Packt:

Django 3 By Example - Third Edition
Antonio Melé

ISBN: 978-1-83898-195-2

- Build real-world web applications
- Learn Django essentials, including models, views, ORM, templates, URLs, forms, and authentication
- Implement advanced features such as custom model fields, custom template tags, cache, middleware, localization, and more
- Create complex functionalities, such as AJAX interactions, social authentication, a full-text search engine, a payment system, a CMS, a RESTful API, and more
- Integrate other technologies, including Redis, Celery, RabbitMQ, PostgreSQL, and Channels, into your projects
- Deploy Django projects in production using NGINX, uWSGI, and Daphne

40 Algorithms Every Programmer Should Know
Imran Ahmad

ISBN: 978-1-78980-121-7

- Explore existing data structures and algorithms found in Python libraries
- Implement graph algorithms for fraud detection using network analysis
- Work with machine learning algorithms to cluster similar tweets and process Twitter data in real time
- Predict the weather using supervised learning algorithms
- Use neural networks for object detection
- Create a recommendation engine that suggests relevant movies to subscribers
- Implement foolproof security using symmetric and asymmetric encryption on Google Cloud Platform (GCP)

Packt is searching for authors like you

If you're interested in becoming an author for Packt, please visit `authors.packtpub.com` and apply today. We have worked with thousands of developers and tech professionals, just like you, to help them share their insight with the global tech community. You can make a general application, apply for a specific hot topic that we are recruiting an author for, or submit your own idea.

Leave a review - let other readers know what you think

Please share your thoughts on this book with others by leaving a review on the site that you bought it from. If you purchased the book from Amazon, please leave us an honest review on this book's Amazon page. This is vital so that other potential readers can see and use your unbiased opinion to make purchasing decisions, we can understand what our customers think about our products, and our authors can see your feedback on the title that they have worked with Packt to create. It will only take a few minutes of your time, but is valuable to other potential customers, our authors, and Packt. Thank you!

Index